Strategic Equilibrium for Cooperative Games: Solutions and applications

Gabriel J. Turbay Bernal, Ph.D.
Giovanni E. Reyes Ortiz, Ph.D.

Strategic Equilibrium for Cooperative Games:
Solutions and applications

Gabriel J. Turbay Bernal, Ph.D.
Giovanni E. Reyes Ortiz, Ph.D.

STRATEGIC EQUILIBRIUM FOR COOPERATIVE GAMES:
SOLUTIONS AND APPLICATIONS

All rights reserve. No part of this book may be reproduced in any form or by mechanical means including mimeograph, tape recorder or any magnetic media without direct permission from the publisher

First Edition

© 2021 by Gabriel J. Turbay Bernal, PhD. (†)

© 2021 by Giovanni E. Reyes Ortiz, PhD

INDEPENDENTLY PUBLISHED

ISBN: 979-854-409-478-4

Design & Layout

Delys Palacios, Esp

Publicaciones SKP Consultores SAS

Review and adaptation:

Félix Socorro, PhD

CEO & CBO

Publicaciones SKP Consultores SAS

Available on Amazon.com

CONTENT

PREFACE 9

INTRODUCTION 11

Towards the Complete Solution of the n-Person Cooperative game and its applications 15

    1. The n-Person Cooperative Game in Characteristic Function Form 22
        1.1 Modeling Cooperation Scenarios 22
        1.2 Von Neumann and Morgenstern Stable Set Solutions 48
        1.3 Genesis of a strategic-equilibrium for the game 65
    2. Auction Equilibrium in Coalition Formation 83
        2.1 Bidding for player 1's cooperation 85
        2.2 Bidding for player 2's cooperation 90
        2.3 Bidding for player 3's cooperation 94
        2.4 Auction Consolidation Summary 98
    3. Utility Transfer Analysis and Stability in Bargaining Scenarios 115
        3.1 Preliminaries 117
        3.2 Utility transfers in bargaining scenarios 138

Turbay Bernal & Reyes Ortiz

This book is a tribute in memory of
Gabriel J. Turbay Bernal, **Ph.D.**
(1947-2020)
for his lifelong contributions, teachings, and achievements.

Turbay Bernal & Reyes Ortiz

PREFACE

The main purpose of this book is to introduce a theory of solutions for the n-person cooperative game through the simple case with 3-persons. It is intended to give the necessary background for readers, students and researchers in the quantitative and social sciences to enhance their theories and approaches with basic mathematical tools applied and developed for game theory analysis within a systems perspective.

Von Neumann and Morgenstern introduced the theory of games as the "proper instrument with which to develop a theory of economic behavior". The snowballing development of game theory and its applications, in the last seventy years, has proven to be not only the proper instrument to develop a theory of economic behavior but the appropriate one for developing the theories for different types of interactive behavior as studied in political, social, environmental, biological, economic and behavioral sciences. Modeling examples of such applications are presented throughout the book.

The study of cooperative games in von Neumann and Morgenstern's *Theory of Games and Economic Behavior* (1947) was presented through cases that gradually increased the number of players. Especial attention was devoted to the 3-person game where most of the essential features of n-person cooperative games can be appreciated and learn from. To focus on such especial case, we consider a solid intuitive model easy to relate to and work with in advancing towards obtaining a complete solution and understanding of the general n-person cooperative game with transferable utility and its applications.

The book is intended for readers with a minimal training in basic finite mathematics, set theory and probability theory. Some mathematical considerations are provided in some of these topics, not necessarily elementary, covering fundamental concepts and theorems used in the theory and applications developed in the book. The main aim here was to complement and keep the presentation of the subject as simple as possible while preserving mathematical rigor.

INTRODUCTION

Game theory is formally a branch of mathematics and economics that models rational behavior in scenarios of conflict and cooperation. It is mainly concerned with the description and properties of interactive goal-oriented behavior and may be considered as a multi-person decision science. A game is considered to be cooperative if binding agreements among the players is possible. Otherwise, is referred to as non-cooperative. Clearly, any non-cooperative game may become a cooperative one by allowing the players, whenever possible, to coordinate strategies and form coalitions based on binding agreements.

In fact, the characteristic function form representation of a cooperative game from was originally obtained in von Neumann and Morgenstern (1944) from the normal form representation of a game in which the outcomes are described in terms of the strategies of individual players. A main concern in the theory to find a satisfactory solution interpreted as standards of behavior for any given game.

In this book we present a theory of solutions for the n-person cooperative game with transferable utility through the simple tractable case of the three-person game. In doing so we address fundamental questions in the theory games of strategy[1] such as: which coalitions are likely to form if the grand coalition doesn't form? And if they form, how will the proceeds of the coalition be to be divided among the players?

Such questions among several do not have a satisfactory answer in the current stage of development of the theory. Here we identify and introduced a fundamental equilibrium, shown to exists for every cooperative game We consider this equilibrium to be a key element in finding reasonable answers to the above questions. Based on the fundamental equilibrium we present our theory of solutions side by side with the two

---

[1] «Games of strategy» is the term given by von Neumann to the games in consideration.

most accepted solution concepts namely the core and the Shapley value We will also relate the proposed solutions to the stable sets also known as von Neumann and Morgenstern solutions in an attempt to explore the immense systemic wealth inherent in these concepts.

The book is organized in three main parts. Part I is mostly concerned with the theoretical aspect of the theory as a model of rational behavior; it includes the basic concepts and definitions, and theoretical elements, such as the core, the Shapley value and the von Neumann and Morgenstern solutions. Part II deals with the interaction of both theory and the modeling of specific applications, and the introduction of the fundamental equilibrium obtained as solution to an auction process. Part III is devoted to issues related to utility transfer analysis and stability in bargaining scenarios.

Towards the Complete Solution of the n-Person Cooperative game and its applications

Social, economic, political and ecological happenings may be perceived as events imbedded in an endless interrelated web of decisions. That is, decisions that are made in response to other decisions. The nature of this web or decision network makes the theory of games of strategy the proper instrument to model interactive rational behavior since it may focus on specific subsets of any wide interactive network in the form of interrelated decision trees. These trees are known as the extensive form of a game. Thus, specific archetypes of these games can be studied in isolation as it has been the case with the three-person simple majority, the non-cooperative prisoner's dilemma or the pure bargaining games. Thus, game theory provides a comprehensive systemic framework as well as a powerful tool for studying interactive human behavior by focusing on specific but well-defined behavioral scenarios.

As defined by von Neumann and Morgenstern (1944), a game is a conceptual unit, whose basic constituent elements are *moves* executed by players or by chance according to certain rules that define the game. A move is an occasion to make a choice among alternatives and these moves occur in chronological order that may be sequential, simultaneous or both.

The game initiates generally with a move made by chance or by one of the players but it may also begin with a set of simultaneous moves made by two or more players. A game terminates likewise with one or several simultaneous moves among a set of possible final moves. Associated with every terminal move there is a payoff for each and every one of the players. The payoffs to the players are expressed in terms of a numerical utility that reflects the true preference over the outcomes and satisfies certain conditions so that players that desire to maximize the utility obtained can do so in conditions of uncertainty by maximizing the expected value of the utility to be obtained. In using decision trees to represent games, special treatment has to be given to the cases involving simultaneous moves.

Any game represented by moves (*extensive form*) can be transformed into an equivalent one described in terms of strategies (*normal form*) and when cooperation can take place, coalitions may form and both strategies and payoffs can be pooled together, the game under consideration can be transformed into an equivalent one in coalitional value form (*characteristic function form*). To pool the payoffs requires the associated utility to be freely transferable and divisible among the players. An example of a game presented in these three equivalent forms is given in Appendix 1. Latter developments of the theory of games after it was presented in von Neumann and Morgenstern (1944) extended these basic game forms introducing the *partition function form* and the *non-transferable utility characteristic function form*.

Here we will approach the problem of solving the n-person cooperative game in characteristic function form with transferable utility. We intend to provide an analytical constructive framework within a systems perspective that allow us to make explicit the equilibriums imbedded in the general n-person cooperative game and that lends itself amenable for applications in the social and behavioral sciences. These hidden equilibriums structural and strategic in nature haven't been fully identified nor mathematically characterized by the main body of research in the theory of games of strategy in the seventy years elapsed after the appearance of the Theory of Games and Economic Behavior. A significant part of research efforts in cooperative game theory have been placed in solution concepts that assume the formation of the grand coalition. Whenever the game is *super-additive*, we may reasonably expect the coalition of all players to be the one likely to form. The final payoffs to the players are consequently in the form of *imputations*. That is, an efficient distribution of the total value to be obtained if the grand coalition forms that assigns to each player an amount not inferior to what each player can obtained by himself. Thus, the formation of the grand coalition has been generally assumed not deduced. Such type of assumption is of intuitive inductive rather than objective deductive nature. Inductive intuition modeling often provides the foundations of axiomatic approaches while objective deduction provides the basis of constructive ones. These two approaches may have synergic complementation especially when the constructive approach is framed within a systems perspective. Such combination of approaches as the one revealed in von Neumann and Morgenstern's book *Theory of Games and Economic Behavior* (1944) have a higher explicative capacity than latter developments in the theory of games which lack the requisite variety of the systems view and have led the theory into unsolved paradoxes and conceptual death ends.

Axiomatic approaches by themselves don't create new knowledge. Neither, as the mathematical proofs of *reductio ad absurdum*: proof by contradiction, throw any light or clue into how one may arrive to the obtained results. This is not to say that axiomatics don't constitutes a powerful method for organizing intuitive reasoning into a well define set of hypotheses where from the main results of a theory can be deduced, organizing new results and consolidating existing ones. But we want to keep in mind that the main purpose of game theory is to shed light into the principles behind explaining interactive rational behavior and not necessarily the proving of theorems per se. When such approaches are complemented with constructive ones, and *vice versa*, the understanding of the subject under consideration is greatly enhanced. The constructive approaches are among the main elements that differentiate between normative and positive views in economics and game theory. The normative view prescribes while the positive view describes. The possible formation of coalition structures in game theory, not necessarily the grand coalition, as in the theories of the kernel and the Bargaining sets, also falls in the category of results that have been assumed not deduced.

Here, we will depart considerably from these and other previous solution concepts by introducing a constructive procedure within a systems perspective. We intend to explain which coalitions are likely to form, previous to considering the formation of the grand coalition. Also, we will identify the possible distribution agreements that explain why such coalitions that may form. In fact, we will extend our analysis traditionally limited to *essential games* to a more general class of games that are essential in a wider sense. This class contains the essential games as a sub-class and will be referred to as the class of *extended-essential games*. We think that the processes of coalition formation and payoff determination must be treated in a concurrent interrelated form. As mentioned above, most of the well-known solution concepts in cooperative game theory impose stability conditions on distributions independently on the coalition formation process which is generally assumed. The conceptual and methodological departure from existing approaches and solution concepts taken here will allow us to better understand the systemic interrelation and role that bargaining alternatives may play, in determining the bargaining power of the players and in defining the coalitions most likely to form in a stage previous to the possible formation of the grand coalition. It will also help us to determine the manner in which the corresponding proceeds, of the coalitions that may form, are to be divided among its members. One would expect rational players to have the means of knowing ¿how well they may do in the game before considering joining the grand coalition? Then such information can be used to bargain and make strategic moves to justify the acceptance or to induce a given distribution of the proceeds of the grand coalition if the grand coalition is to form.

1. The n-Person Cooperative Game in Characteristic Function Form

Mathematical models are descriptions of systems of interrelated elements. These systems are mental scenarios constructed by an observer of presumably existing objective or imaginary realities. Based on some fundamental assumptions, these models are intended to characterize the properties, explain the functioning and predict the behavior of the system in consideration by means of equations and logical propositions. Here, to model cooperation, we will take von Neumann and Morgenstern's intuitive interpretation of the cooperative game as a social economy which will leads us directly to the description of the game in characteristic function form without resorting to the normal form.

1.1 Modeling Cooperation Scenarios

A cooperation scenario is one in which a group of individuals with parallel interest and common purpose, voluntarily act together to achieve an end of mutual benefit. The scenario we want to consider here for mathematical modeling is one where there are two or more persons, n in number. Each person can create some value for himself. However, instead of being by themselves, the persons under consideration can freely choose to cooperate in groups that presumably have a higher per capita value creation capability. Whenever such is the case, we may say there is an incentive to cooperate, given that the value generated by the group is higher than the sum of the individually created values. We would like to think that rational individuals prefer to obtain the highest possible share of the cooperation benefits and that they will choose to act accordingly so as to maximize the value obtained for them under conditions of certainty or the expected value if uncertainty is present with known probabilities. Here, each of the n persons will join one and only one of the groups that may form so that no two groups have a person in common. Clearly, every person will end up being either in a group with two or more but less than or equal to n persons, or alone with himself. It is assumed that the value generated by a group when it forms into a coalition can be unrestrictedly redistributed among the members of the group. The archetypical scenario above describing a social economy is known as the *n-person cooperative game with transferable utility*. The persons involved are referred to as *players* and the groups that may form are actually subsets of the set of n players and referred to as *coalitions*. The value that the players and the coalitions can create is a numerical *utility* which for modeling purposes, here will be assumed to be a transferable commodity that such as money. However, the term utility will be used in general unless otherwise specified. The functional correspondence of utility values to coalitions is the characteristic function.

In game theory, the players are distinguished by a number in the set N = {1, 2, ..., n}. These players are assumed to act rationally. That is, they will analyze, anticipate and take into account the combined effect of his individual moves with the possible moves of the other participants in defining the final outcomes and they will act accordingly so as to maximize the utility they may obtain. Thus, game theory seeks to find the mathematical principles which define rational behavior.

Here we will begin our exposition by giving special attention to the transferable utility three–person cooperative game. In this case n = 3 and each player j, j = 1, 2, 3 in the set N={1, 2, 3} can secure by himself an amount of utility equal to v_j. By coordinating join efforts, any of the possible subsets of players: {1, 2}, {1, 3}, {2, 3} and {1, 2, 3} can form as a coalition and together obtain the corresponding amounts v_{12}, v_{13}, v_{23} and v_{123}, respectively. Given that a player can belong to one and only one coalition, for n = 3 there are only 5 possible final outcomes[2] that consist on partitions of the set of N players or groups of coalitions that might form and are known as *coalition structures*. The parametric description of these structures is given in the list below that includes the corresponding values to be obtained by the coalitions in each coalition structure:

$$
\begin{array}{lll}
CS1: & \{1\}, \{2\}, \{3\}; & v_1, v_2, v_3 \\
CS2: & \{1, 2\}, \{3\}; & v_{12}, v_3, \\
CS3: & \{1, 3\}, \{2\}; & v_{13}, v_2 \\
CS4: & \{2, 3\}, \{1\}; & v_{23}, v_1 \\
CS5: & \{1, 2, 3\}; & v_{123}
\end{array}
\qquad (1.1)
$$

---

[2] The number of partitions of a set are given by the exponential Bell numbers B(n). For n =3, B(3) = 5. The corresponding number of partitions for n = 20 is given by the astronomical number B(20) = 51724158235372. (Approx. 5.17 trillions) See Appendix A7.

If cooperation in forming a coalition is to be considered by the players and one of the above coalition structures occurs, it is reasonable to expect the player's willingness to cooperate to be conditioned to the amount of utility that each player is assured to obtain once a particular coalition is formed. The enforceability of these agreements is a fundamental assumption made in modeling cooperation. It is often single out as a major characteristic to distinguish cooperative from non-cooperative games even though agreements may occur in the later ones. Agreements, here will be assumed to be binding.

Let x_j denote the amount that player j receives when a given coalition is formed and the proceeds of the coalition are divided among its members. The value x_j is referred to as the *payoff* to player j. The set of payoffs to the players, one for each player, can be given in the form of a *payoff vector* $x = (x_1, x_2, \ldots x_n)$ *in n-Dimensional Euclidean Space R^n*. Thus, a payoff vector could be any vector in n-dimensional Euclidean space. Without loss of generality, we will concentrate on non-negative payoff vectors which are elements in the non-negative orthant of

R^n denoted by R^n_+. We may expect such allocation of the value of the coalition to meet certain rationality conditions or principles if cooperation is to take place. Say that coalition $\{1, 2, 3\}$ is to be formed and the corresponding payoffs x_1, x_2 and x_3 are being considered by the players. They certainly would like to check if these satisfy the following conditions:

$x_1 \geq v_1$, $x_2 \geq v_2$ and $x_3 \geq v_3$

(individual rationality) (1.2)

$x_1 + x_2 \geq v_{12}$, $x_1 + x_3 \geq v_{13}$ and $x_2 + x_3 \geq v_{23}$

(coalitional rationality) (1.3)

$x_1 + x_2 + x_3 = v_{123}$

(group rationality) (1.4)

The first condition (1.2) states the individually rational requirement that the amount x_j that player j will receive when the coalition forms is to be greater than v_j, namely the value he, player j, can obtain by himself. Otherwise, there would be no point in cooperating. The second condition (1.3) states the conditions that the proposed payoffs to the players must be such that their sum for any particular coalition must be no less than the value that such coalition can obtain by forming an independent group; otherwise, they would rather consider joining together to obtain a higher overall value. The group rationality condition (1.4) is one of efficiency. If the coalition of all players is to form, the total amount to be distributed need not be less, nor can be more, than the total value that can be obtained.

Remark 1.1 The coalition rationality condition (1.3) is a reasonable requirement for proposed payoffs in the coalition formation process but not one to describe the final outcomes of cooperation. In the processes of coalition formation some coalitions might not be able to be formed if some of their players have previous commitments with the formation of other coalitions. And hence it is very possible to have, as final outcomes, imputations that are not coalitionally rational.

The coalition of all players $N = \{1, 2, \ldots, n\}$ under consideration is referred to as the **grand coalition**, aside from the rationality conditions which the distribution of the proceeds of a coalition must satisfy, we observe that for the players to consider the formation of any given coalition, before reaching an agreement on how to divide the proceeds, the values that the coalition can secure for itself in a game with n=3 must satisfy the following conditions:

$v_{123} > v_1 + v_2 + v_3$ for the grand coalition (essentiality) (1,5)

$v_{ij} > v_i + v_j$ for any two-person coalition $\{i, j\}$, $i, j = 1, 2, 3$; $i \neq j$ (1.6)

That is a player will not even consider joining a coalition if the value which the coalition can obtain when it forms is not greater than the sum of the amounts that the players can get by themselves. Clearly, in such situation at least one player would get less than what he could secure by himself.

Note the relation that exists between a coalition and the value which the coalition may obtain when it forms is a functional one. Such type of relation is mathematically known as a real valued set function and in game theory is referred to as **the characteristic function** of the game, Thus the elements of the n-person cooperative game are: A set of $N=\{1,2, ..., n\}$ of rational players and a real valued set function $v: 2^N \rightarrow R$ defined on the 2^n subsets $S \subseteq N$ with

$v(\emptyset) = 0$. (Se appendix A2 on set theory and convexity) Thus the characteristic function for our three-person game in parametric presentation is given by:

$v(\emptyset) = v_0 = 0$, $v(\{1\}) = v_1$, $v(\{2\}) = v_2$, $v(\{3\}) = v_3$,

$v(\{1, 2\}) = v_{12}$, $v(\{1, 3\}) = v_{13}$, $v(\{2, 3\}) = v_{23}$ and $v(\{1, 2, 3\}) = v_{123}$,   (1.7)

The characteristic function values of the coalitions in N are the parameters $v_0, v_1, v_2,..., v_{123}$. These are real numbers representing amounts of transferable utility. Here, the utilities are assumed to be money or money-like utility, to make this presentation simpler[3]. Formally we have

Definition 1.1 An n-person cooperative game in characteristic function for is a pair $\Gamma = (N, v)$ where $N=\{1, 2, ...n\}$ is the set of players and $v: 2^N \rightarrow R$ is a real valued set function defined on the subsets of N with $v(\emptyset) = 0$.

Definition 1.2 An n-person cooperative game $\Gamma = (N, v)$ is said to be a *transferable utility* game if the characteristic

---

[3] For a more precise description of the role of utility in game theory see appendix A3.

function values of the corresponding coalitions can be freely distributed among its members.

Every n-person cooperative game considered here will be assumed to be a transferable utility one.

Definition 1.3 A set valued function v: $2^N \to R$ is said to be *super-additive* if

$$v(S \cup T) \geq v(S) + v(T) \text{ whenever } S \cap T = \emptyset \; ; \; S, T \subseteq N$$
(super-additivity)   (1.8)

Unless otherwise specified we will assume the games here to have super-additive characteristic functions. The supper-additive property captures the synergic characteristic of creative cooperation: "The whole is greater than the sum of its parts". There is strength in union". Clearly, if the game is super-additive, the players may have an incentive to form. Games of interest to us here are those where the players may benefit from forming a coalition. The characteristic function v of a game $\Gamma = (N, v)$ is often used to refer to the game itself. Thus, a game is said to be super-additive if the characteristic function of the game is so. Whenever the game is supper-additive and the formation of the grand coalition is assumed, for the game to be non-trivial the essentiality condition (1.4) must hold for the coalition of all players.

Definition 1.4 A cooperative game Γ= (N, v) is sad to be essential if and only if

$$v(N) > \sum_{j=1}^{N} v_j \qquad (1.9)$$

We will be working with cooperation scenarios in a stage previous to considering the formation of the grand coalition where the players must take into account what they may obtain in all possibilities conditions. To describe the simultaneous possibilities a player may have we need to consider collections of subsets of N in the form of a cover of N which includes the concept of partition as a special case. *A cover of N* is a collection of subsets of the set N whose union gives N.

Definition 1.5 A cooperative game Γ= (N, v) is sad to be extended-essential if and only if

$$v(C) > \sum_{j \in C} v_j \; for \; all \; C \; in \; a \; cover \; \mathbb{C} \; of \; N \qquad (1.10)$$

Remark 1.2 Note that essential game case is but a special e-essentiality one, namely when de cover $\mathbb{C}$ of N is equal to N itself.

The strategic possibilities available to the players in a game remain invariant if there is a proportional variation in the characteristic function values of the coalitions or if such values are modified by adding (or subtracting) fixed amounts for each player in every coalition. The modified characteristic function gives us a game which is *strategically equivalent* to the original one.

Definition 1.6 Two games $\Gamma_1 = (N, v)$ and $\Gamma_2 = (N, u)$ are said to be strategically equivalent if and only if there exist real numbers $\alpha_0, \alpha_1,...,\alpha_n$, with $\alpha_0 > 0$ such that:

$$v(S) = \alpha_0 u(S) + \sum_{j \in S} \alpha_j \qquad (1.11)$$

Example 1.1 Let N={1, 2, 3}, a) the game $\Gamma_1 = (N, u)$ be given by u({1}) = 20, u({2}) = 15, u({3}) = 5, u({1,2}) = 85, u({1,3}) = 75, u({2,3}) = 60, u({1,2,3}) = 100. And b) the game $\Gamma_2 = (N, v)$ be given by v({1}) = 0, v({2}) = 0, v({3}) = 0, v({1,2}) = 70, v({1,3}) = 50, v({2,3}) = 40, v({1, 2,3}) = 100 We may verify that the games Γ_1 and Γ_2 are strategically equivalent with real constants $\alpha_0 = 1$, $\alpha_1 = 20$, $\alpha_2 = 15$, $\alpha_3 = 5$. If we let c) $\Gamma_3 = (N, w)$ where w({1}) =0, w({2}) =0, w({3}) =0, w({1,2}) = 0.7, w({1,3}) = 0.5, w({2,3}) =0.4, w({1, 2,3}) = 1, we may check that the games given in a), b) and c) above are all strategically equivalent to each other.

Remark 1.3 Strategic equivalence is an *equivalence relation* among the set of all n-person cooperative games. In mathematics, such relations are characterized by being i) reflexive, ii) symmetric and iii) transitive. (See Appendix Ax)

Definition 1.6 An n-person cooperative game is said to be in $(0,\eta)$ - normalization if and only if

$$v(N) = \eta \text{ and } v(\{j\}) = 0 \text{ for } j = 1, ..., n \qquad (1.12)$$

Proposition 1.1 Every n-person cooperative game $\Gamma'=(N,u)$ is strategically equivalent to a unique game $\Gamma = (N, v)$ in $(0, \eta)$ –normalization.

Remark 1.4 In example 1.1 the game Γ_2 in b) is in (0,100)-normalization and the game Γ_3 is in (0,1)-normalization. Without loss of generality, through this book we will work with games that are in (0,100) normalized form. So that our generic characteristic function in parametric form for the three-person cooperative game in (1,6) becomes

$v(\emptyset) = 0$, $v(\{1\}) = 0$, $v(\{2\}) = 0$, $v(\{3\}) = 0$,

$v(\{1, 2\}) = v_{12}$, $v(\{1, 3\}) = v_{13}$, $v(\{2, 3\}) = v_{23}$ and

$v(\{1, 2, 3\} = 100$, (1.13)

If the coalition N of all players form, the distribution of its value v(N) among its players in payoffs which satisfy the individual rationality condition as in (1.2) is known as an *imputation*.

Definition 1.7 A payoff vector $x = (x_1, \ldots, x_n)$ in R^n is said to be an imputation if and only if

$$\sum_{j \in N} x_j = v(N) \text{ and } x_j \geq v_j \geq 0,$$

$$j = 1, \ldots, n \quad (1.13)$$

The set of all imputations for a given (0, η)-normalized game Γ = (N, v) (see figure 1.1 below) consist of an **n-1 simplex** I, with η = v(N),

$I = \{x \in R_+^n \mid \sum_{j \in N} x_j = v(N), j = 1, \ldots, n\}$ \quad (1,14)

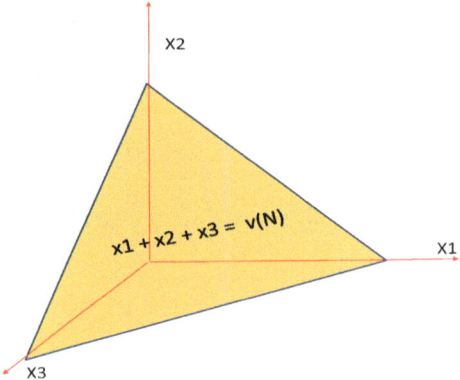

Figure 1.1 Set of Imputations for the n-person cooperative game with n = 3

The set of imputations of an n-person cooperative game is a special type of polyhedral compact set known as a n-1 dimensional simplex described equivalently by a system of inequalities or as the convex hull on n points affinely independent (see Appendices X and Y).

Thus, an imputation is a distribution of the characteristic function value corresponding to the grand coalition that satisfies individual rationality (1.2) and efficiency (1.4).

Cleary, if we can find an imputation that satisfy coalitional rationality conditions in (1.3) then such imputation if proposed to the players would have a certain degree of stability around which cooperation can take place. This is so because no player or group of players can all improve upon to obtain a higher value by joining in an alternative coalition. The set of payoff vectors that satisfy all three rationality conditions (1.3), (1,4) and (1.5) simultaneously, is known as *the core* of the game and here for the n-person cooperative game is formally defined as

Definition 1.8 An imputation x is in the core $C(\Gamma)$ of the game $\Gamma=(N,v)$ if and only if

$$C(\Gamma) = \{x \in R^n \mid x(S) \geq v(S) \text{ for all } S \text{ in } N \text{ and } x(N) = v(N) \} \qquad (1.14)$$

Example 1.2 For the (0,100)-normalized game $\Gamma=(N,v)$ with characteristic function $v(\{j\})= 0$ for $j=1,2,3$; $v(\{1,2\} = 80$, $v(\{1,3\}) = 60$, $v(\{2,3\}) = 40$, and $v(\{1,2,3\}) = 100$, the core is given by the set $C(\Gamma)$ of imputations $x = (x_1, x_2, x_3)$ that satisfy the conditions

$$x_1 + x_2 \geq 80, \quad x_1 + x_3 \geq 60 \text{ and } x_2 + x_3 \geq 40 \qquad (1.15)$$

The core $C(\Gamma)$ of the above game is given in figure 1.2 below. The first graph provides us with a 3-Dimensional view of the set of imputations with the core as a subset constraint by the coalitional rationality conditions in (1.11).

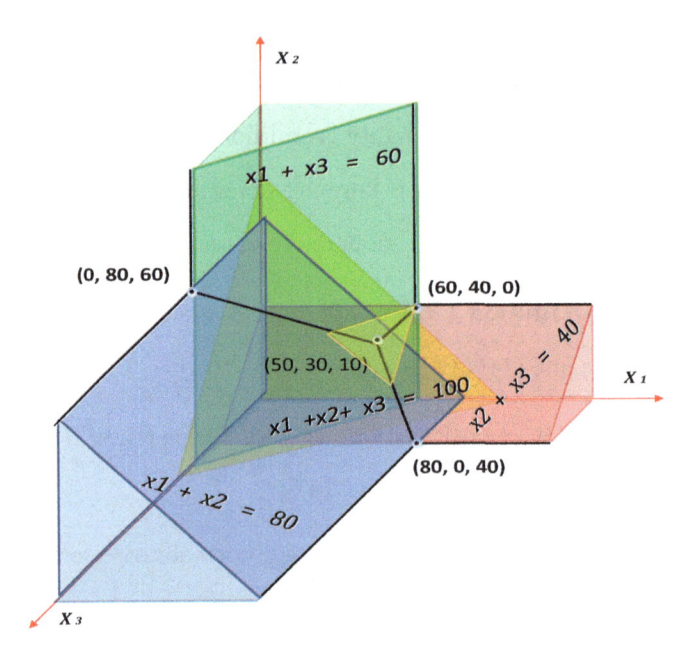

Figure 1.2 Three–Dimensional view of the core C(Γ) of a 3-person game

By focusing on the imputations set, we obtain a view of the core in two dimensions as a subset constraint by the intersection lines of the coalitional rationality constraint planes with the with the imputations set.

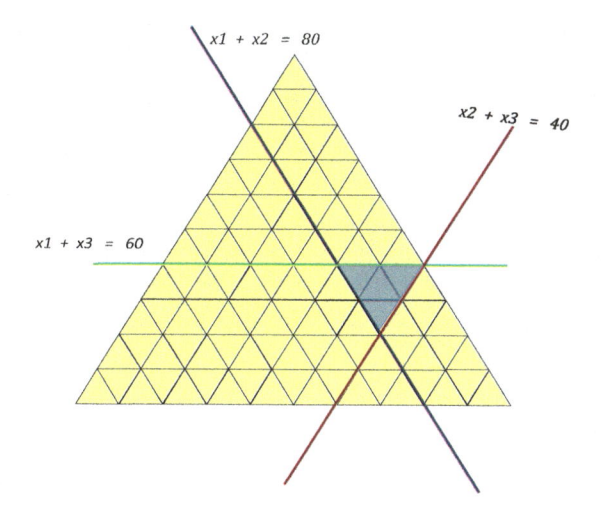

Figure 1.3 Two–Dimensional view of the core C(Γ) of a 3-person game

Sets in R^n defined by linear equations and linear inequalities are known as *convex polyhedral sets*, Convex sets are those for which given any two points in the set, all other points in the line segment joining them are in the set. The set of points in R^3 that satisfy the individual and coalitional rationality conditions (1. 2) and (1.3) constitute a convex polyhedral set. We can observe in figure 1.2 of example 1.3 such set consisting of all points that are above the planes defined by the equations corresponding to the inequalities in (1.11) intersecting at the point (50, 30,

10). Such points are not necessarily imputations and we will be referred to as *coalitionally rational extended imputations*.

Definition 1.9 An extended imputation is a payoff vector $x = (x_1, \ldots, x_n)$ that satisfies individual rationality. That is,

$$x_j \geq v_j \text{ for } j = 1, \ldots, n \qquad (1,16)$$

Therefore, a coalitionally rational extended imputation is one that satisfies the coalitional rationality condition

$$\sum_{j \in N} x_j \geq v(S) \quad for\ all\ \ S \subset N \qquad (1.17)$$

The *value level* of an extended imputation relative to a coalition S in N is defined as

$$x(S) = \sum_{j \in S} x_j \quad for\ all\ \ S \subseteq N \qquad (1.18)$$

Remark 1.5 For 0-normalized cooperative games all individual player's characteristic function value is equal to 0 and hence the set of extended imputations is the set R^n_+. That is, the set of non-negative vectors in R^n.

Definition 1.10 An extended imputation is said to be *detached* if and only if

$$x(S) = \sum_{j \in S} x_j \geq v(S) \quad \text{for all } S \subseteq N \quad (1.19)$$

The core $C(\Gamma)$ consists of the intersection of the set of coalitionally rational extended imputations with the imputations simplex. That is, the intersection of two convex polyhedral sets which always happens to be a convex polyhedral set. It follows then that

Proposition 1.2 The core of an n-person cooperative game is a compact[4] convex polyhedral set.

Remark 1.6 As mentioned above, the core of a game is clearly a set of possible distributions of the value of the grand coalition around which cooperation might take place. However, it is not necessarily a good predictor of the possible outcomes that would result from rational behavior in bargaining for cooperation. In fact, in wide classes of games, among them the class of *zero-sum games*, and the one of *constant sum games*, the core is empty. In

---

[4] A compact set is one that is closed (contains all its boundary points) and bounded (contained in an n-dimensional circle with finite radius)

many games the core is too large. Such is the case with *pure bargaining games* where the core coincides with the imputations set. Pure bargaining games are those in which the characteristic function value of all coalitions other than the grand coalition is zero. Formally we have,

Definition 1.11 An n-person cooperative game $\Gamma = (N, v)$ is zero-sum if and only if

$$v(S) + v(N/S) = 0 \text{ for all } S \subseteq N \quad (1.20)$$

Definition 1.12 An n-person cooperative game $\Gamma = (N, v)$ is constant sum if and only if

$$V(S) + v(N/S) = k \text{ for all } S \subseteq N, \text{ k a real constant} \quad (1.21)$$

Proposition 1.3 Every zero-sum game is strategically equivalent to a constant-sum game

Definition 1.13 An n-person cooperative game $\Gamma = (N, v)$ is a pure bargaining game if and only if

$$v(S) = 0 \text{ for } S \subset N, \text{ and } v(N) = k, k \neq 0 \text{ a real constant} \quad (1.22)$$

Proposition 1.4 Every constant sum game has an empty core.

Example 1.3 The three-person game $\Gamma = (N, v)$ with characteristic function given by $v(\{1\}) = v(\{2\}) = v(\{3\}) = 0$, and $v(\{1, 2\}) = v(\{1, 3\}) = v(\{2, 3\}) = v(\{1, 2, 3\}) = 100$ is clearly constant sum since condition (1.13) is satisfied: the sum of the value of any coalition S with the value of its complement to N, namely N/S is the real constant 100. This game models the archetypal scenario known as the *simple majority game*. In this game any group of players that constitute a simple majority can win a totality represented 100 (percent).

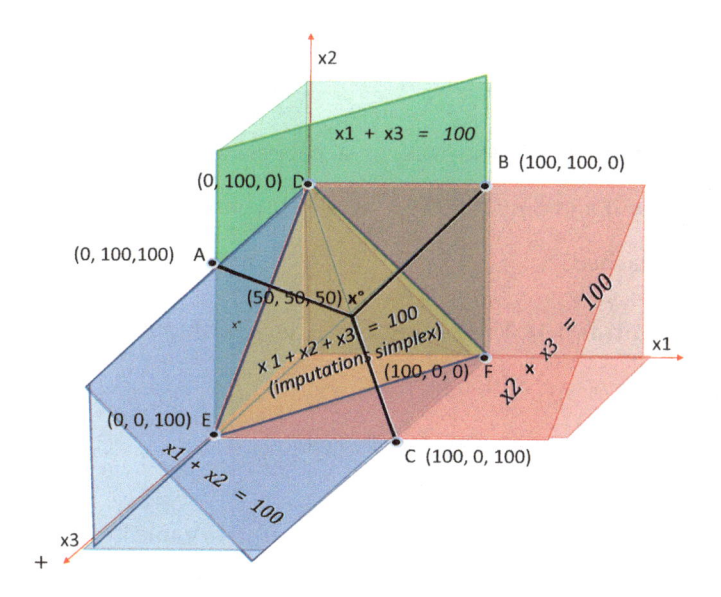

Figure 1.4 Imputations and Rationality constraints in simple majority game

It is easy to see in figure 1.4 that there is no intersection between the set of coalition rational extended imputations and the set of imputations. Also, we may observe that the value level of any imputation x is x(N) =100 and that the value level of any coalitionally rational extended imputation has to be greater than or equal to x°(N) =150. Thus, there is no imputation that can satisfy all rationality conditions. This clearly shows that for the 3-person simple majority game the core is empty. That is C(Γ) = ∅.

Proposition 1.5 The core of a cooperative game is not empty if and only if there is at least one imputation which is detached

Definition 1.14 The excess of an extended imputation x in R relative to coalition S:

$$e(x, S) = \sum_{j \in S} x_j - v(S), \quad \text{for all } S \subseteq N \tag{1.23}$$

Proposition 1.6 Every n-person cooperative game Γ= (N, v) determines a number |Γ| with the property that a detached extended x with e(x, N) = e exist[5] if and only if

---

[5] This proposition was first introduced and proof in vN-M (1944/1957) pp.370-371

$$e \geq |\Gamma| \quad (1.24)$$

Here the number $|\Gamma|$ will be referred to as the *von Neumann number* of the cooperative game.

Clearly, for any imputation the excess $e(x, N) = 0$. Let $x°$ be any extended imputation for which $e(x°, N) = |\Gamma|$ then a detached imputation with $e=0$ exists if and only if $e = 0 \geq e(x°, N) = e = 0 \geq e(x°, N) = |\Gamma| = \sum_{j \in N} x_j^o - v(N)$, or equivalently a detached imputation exists if and only if $0 \geq (x°(N) - v(N))$. Thus, as a consequence of proposition 1.6, the non-emptiness of the core in proposition 1.5 becomes:

Proposition 1.7 The von Neumann number of any cooperative game is given by the excess of the extended imputation $x°$ with respect to N

$$|\Gamma| = \sum_{j \in N} x_j^o - v(N) \quad (1.25)$$

Corollary 1.1 The core of a cooperative game is not empty if and only if $0 \geq |\Gamma|$ *or equivalently if* $v(N) \geq x°(N)$.

It is easy to see that the number $|\Gamma|$ associated with every game is determined by the value of the grand coalition v(N) and the value level x°(N). Both a detached imputation x° ant its value level can be determined by means of a linear programing problem that can be associated with every cooperative game and that will be introduced and used in the following sections.

Example 1.4 For the 3-person simple majority game given in example 1.3 the imputation simplex is the set of *convex combinations* of the three points A (0,100,0), B(100,0,0) and C (0, 0,100) . The von Neumann number for this game is $|\Gamma| = 50$. Since the excess of any imputation is e(x,N) = 0 and 0 is not greater than $|\Gamma| = 50$ then the core of the 3-person simple majority game must necessarily be empty.

Up to this point we have most of the basic constituent elements of our cooperative game scenario described in mathematical form. We have exhibited some important interrelations and properties. Now we are prepared to address some fundamental questions that need be answered satisfactorily:

- Which coalitions are likely to form?
- Which coalitions are likely to form if the grand coalition doesn't form?

- Which coalition structures are likely to form? (1.26)
- If a coalition forms, how will the proceeds be to be divided among its players?
- How well can the players do previous to the formation of the grand coalition?

We will continue with our modeling of rational behavior in cooperative scenarios by using a constructive approach to develop a theory of solutions that answers the above fundamental questions for the 3-person cooperative game. Here we assert that these questions have not been satisfactorily answered in the current state of the theory of games. The theory to be developed in what follows generalizes von Neumann and Morgenstern symmetric solution for the 3-person zero-sum game and also opens a window to appreciate and understand the immense systemic wealth and outcome explicative capacity imbedded in vN-M stable sets solution concept.

1.2 Von Neumann and Morgenstern Stable Set Solutions

In optimization theory with one decision maker an optimal solution or "first" element or optimal solution in a given set is usually obtained by selecting, among the variable elements of the set, one or several that renders the maximum value for a certain objective function. That is, the maximum of a function of the variables that characterize the elements of the set. Whenever the characteristics of the elements are not commensurable, when more than one decision-maker is involved or when several objective functions are to be maximized simultaneously by selecting among elements of a set given, such a first element may not exist at all. Any superiority relation among elements doesn't yield a well-ordered set were from a first element can be found. In general, it is always possible to identify a set of elements for which no superior elements exist. Such set is known as the ***Pareto optimal set***, Pareto boundary or Pareto frontier. Von Neumann and Morgenstern made explicit the existence of "optimal sets" that generalize the concept of an optimal solution which include optimal solutions and Pareto optimal sets as special cases. These concepts have received considerable attention in disciplines other than game theory. Here in explaining the concept we will present a few examples of simple applications in areas as diverse as dominant groups in sociology, optimal team selection, industry analysis and integer programming. In all cases presented the elements of the solution constitute an interrelated system of elements where no element by itself can be considered as a solution unless it exists as a first element.

Let us consider the special case of a social economy viewed as a game of strategy and any distribution of the proceeds of the society of n players, given by a utility n-vector, as a social situation. We may say that a distribution x is "superior" to a different distribution y if and only if for every player j in N, $x_j > y_j$. Unfortunately, in the case of imputations no superior elements can be discriminated by the Pareto frontier. If the total proceeds of the society are to be distributed among its members and these satisfy the individual rationality condition (1.2) the distributions in consideration must be imputations and the set of all imputations constitutes the Pareto frontier for the game under the superiority relation among distributions defined above. Clearly If both x and y are imputations neither x is superior to y nor y can be superior to x. Thus, to compare social situations x and y one must specify for whom is better x than y and also if those for which the social situation x is better than y can protect of defend the amounts distributed by acting as an independent group. The possibility of defending a given distribution by a coalition of players allows for a definition of superiority among imputations but restricted to the supporting coalition. This superiority is termed as **domination** and allows for the definition of a solution for games in characteristic function form in terms of un-dominated imputations

Definition 1.15 A distribution vector $x = (x_1, \ldots, x_n)$ is said to be Pareto optimal in a set of possible distributions D if and only if there is no y in D such that $y_j > x_j$ for all j in N

Proposition 1.8 The set of imputations in a cooperative game is a Pareto optimal set individually rational distributions of $v(N)$

Definition 1.16 An imputation x is said to dominate imputation y if there is a coalition $S \subset N$ such that

$$x_j > y_j \text{ for all } j \in S$$

$$\sum_{j \in S} x_j \leq v(S)$$

Definition 1.15 A set of imputations V is a stable set or vN-M solution for the cooperative game if and only if

No y in V is dominated by any x in V   (1–)

Every y not in V is dominated by some x in V   (1.)

Example 1.5 To better understand how vN-M stable sets generalize the concept of optimum, consider a school class consisting of a set S with 15 students, S = {A, B, C, D, E, F, G, H, I, J, K, L, M, N, O}. For each student two performance measures are given: (Q) for quantitative skills and (H) for social skills. We say that student X outperforms student Y denoted by $X \succ Y$ if and only if $Q(X) > Q(Y)$ and $H(X) > H(Y)$. Clearly, for the scores given below, the set V={A, B, C, F, H, O} is a Pareto optimal set which is obtained applying the definition of the vN-M stable set: (1) No student in V outperforms any other student in V and (2)

Any student not in V is outperformed by some student in V. Here the set V constitutes an "elite group" or "dominant class" of students, see figure below. One would expect that as a team the selected *group* would have superior skills to any other group in dealing with related issues that require both quantitative and social skills.

STUDENT SOCIAL AN QUANTITATIVE SKILLS PERFORMANCE MEASURES

Student Skill	A	B	C	D	E	F	G	H	I	J	K	L	M	N	O
Q	94	70	48	67	80	98	55	86	85	60	30	40	55	89	80
H	64	93	97	65	60	40	82	70	45	88	89	70	65	30	82

No single element of V = {A, B, C, F, H, O} can be thought of as a solution or "first element", though every element of V is a first element relative to some proper subset of S. If Instead of the given values for student J we would have Q(J) = 99 and H(J) = 98 then J would outperformed every element of S except himself and in such case V= {J} would be the a vN-M solution consisting of a singleton containing a first element relative to the outperformance relationship. Note that any linear function of Q and H with non-negative coefficients ie: any weighted average of Q and H would attain its maximum value at some point in V. Hence, V is a set of interrelated elements that as a whole constitute an optimal like solution.

52

The procedure above can be readily extended to find a "dominant class" solution for elements characterized with more than 2-variables. For example, the selection of a "core team" of players in next world soccer cup could be initially determined[6] if m competence measures are given for n players of which we want to identify the set of un-dominated players from which one can select an Q "optimal" team where players might complement each other.

Let us consider now the 3-person zero sum game in 0-1 normalized form ie: the simple majority three-person game, where the characteristic function is given by $v(\{i\}) = 0$, $i=1,2,3$ $v(S) = 1$, $S \subseteq N = \{1,2,3\}$. The set of imputations I consists on all possible distributions of a unit among three persons; that is, the set of vectors $x = (x_1, x_2, x_3)$, with $x_i \geq 0$, $i = 1, 2, 3$ and $x_1 + x_2 + x_3 = 1$. This set is a two-dimensional simplex with corresponding graph given bellow:

---

[6] Many sport associations keep quantitative records of a set performance measures considered as good indicators of the player's competences required to win a game.

Imputations simplex for a 3-person game

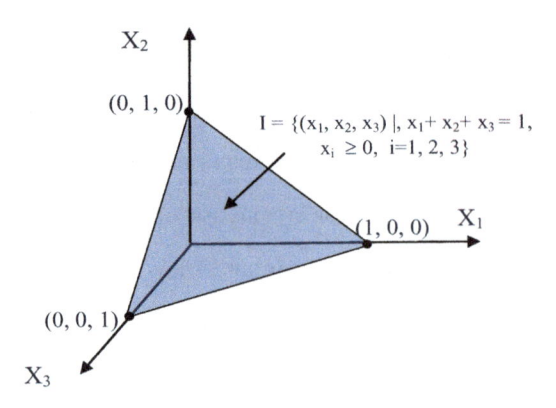

Figure 1.5 *Imputations simplex*: possible allocations of a divisible unit among three persons.

Given that any two players can get together and form a coalition which secures one unit to be divided between them, leaving the excluded player with nothing, we proceed to analyze arbitrary divisions of the unit in question to see how they stand as possible outcomes of "rational behavior".

Let us assume that negotiations may go on until a 2-person coalition form. The issue of the formation of the grand coalition is postponed for now. We assume that a consensus or binding agreement on the division of the proceeds between the members must take place previous to the formation of the coalition. W can readily observe, as explained in vN-M, that only the equal distribution of the proceeds between any 2-players may go unchallenged. These are maximum sustainable claims the players can maintain in all possible circumstances[7] and are given by the set of three imputations:

$V_0 = \{(½, ½, 0), (½, 0, ½) (0, ½, ½,)\}$

These three imputations not only emerge as a unique schema of defensible claims but from the point of view of domination they also possess the following stability vN-M properties (1) No imputation in V_0 dominates any other imputation in V_0 and (2) every imputation in I not in V is dominated by at least one imputation in V. Imputations that satisfy these two properties are presented by vN-M as a solution to the game. So, V_0 is clearly a solution not only

---

[7] For a procedure by which these maximum claims are obtained see vN-M pp. 228-231 and pp.261-262. The type of systemic equilibrium inherent to the 3-person zero-sum game nondiscriminatory solutions is generalized for all cooperative games and characterized and explained in terms of balance collections and theorems of the alternatives for matrices in a following chapter on utility transfers..

in the sense that describes the possible outcomes that rational players will end up with, but mathematically conform as the vN-M generalized optimum when the domination relation is considered to be the optimality criteria. The graph of the solution is given bellow:

Non-discriminatory "objective" solution

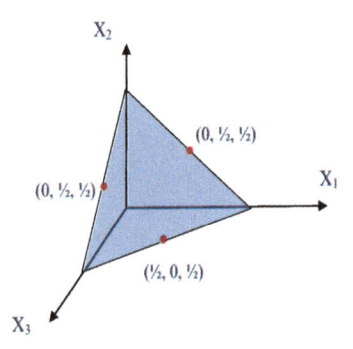

Figure 1.6. *Non-discriminatory "objective solution"*

Given two different imputations x and y the domination relation definition here is based on two types of conditions that must be met for x to dominate y via coalition $S \supset N = \{1, 2, 3\}$ and denoted by $x \succ_s y$:

The preliminary conditions:

C1. S is not empty,

C2. S is effective for x, ie: $v(S) \geq \sum_{j \in S} x_j$.

And the main condition

C3. $x_j > y_j$ for all $j \in S$

Now, in plain words, for x and y different imputations, x dominates y, ($x \succ y$) if there exist a non-empty coalition S where every players in S can get more in x than they are receiving in y and the value the players can collectively secure, if S forms, is greater than or equal to the sum of the amounts that the imputation x assigns to the players in S. The last requirement is to ensure attainability.

Other sets of imputations that satisfy the three vN-M properties above for a solution emerged as necessary consequences of the mathematical concept of solution. These were not preconceived or expected but on the contrary they were required to be explained in terms of rational behavior on how they could come about. These are the solution referred to as sets of the discriminatory type V_j, j=1,2,3. For the specific case where j=2, we have:

$V_2 = \{ (x_1, x_2, x_3) \mid x_1 = a, x_2 = c, x_3 = 1- (c + a), 0 < c < ½, 0 < a \leq 1 - c\}$

Discriminatory solutions to the 3-Person zero-sum game

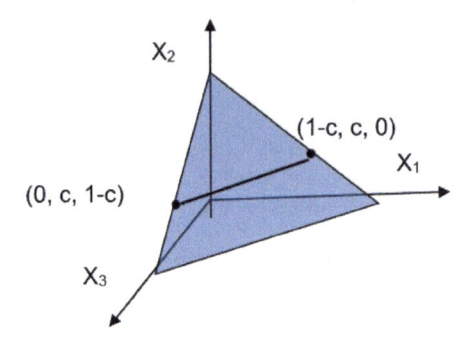

Figure 1.7 *Discriminatory solution, $j = 2$ is the discriminated player ie: $x_2 = c < 1/2$*

Von Neumann and Morgenstern interpretation[8] of these solutions is one where two of the players exclude the third one from the possibility of forming a coalition, assigning him a fixed amount c which he gets in any possible outcome. By doing so, they eliminate the possibility of exclusion for themselves and with certainty they can bargain freely between them the remaining amount 1-c. Their interpretation leaves some points that we think require further explanation in terms of rational behavior both in the 3-person zero sum game as well as in

---

[8] See vN-M pp.288-270 for the zero-sum game, and 418 for the general-sum

the 3-person general sum game. For example, how can we explain outcomes in the solution such as the extreme points (0, c, 1-c) and (1-c, c, 0) if player 2 is excluded from the possibility of participating in the formation of a coalition? Subsequently we will develop the tools that allow us to approach these issues from the perspective of rational behavior with solid and, hopefully (possibly), very convincing arguments on the emergence of these and other possible outcomes difficult to be explain without considering the formation of syndicates beyond the one of coalitions and further the possibility of a player to move, apparently in favor of other players, the strategic equilibrium of the game[9].

An explanation of the "objective" non-discriminatory solution is as follows. V_0 can only be taken as system of interrelated options or possibilities open to the players, where the actual values in one option only make sense in that are related to the other options where the players may protect their specific claims:

---

[9] See below the bargaining alternatives to solve the stronger player paradox

Player			Final Partition	
1 2 3			S	N-S
($x_1°$ $x_2°$ 0) if			{1, 2}	{3} occurs
($x_1°$ 0 $x_3°$) if			{1, 3}	{2} occurs
(0 $x_2°$ $x_3°$) if			{2, 3}	{1} occurs

Here $x_j° = 1/2$, j=1,2,3 are the maximum sustainable claims that can be reasonably maintained by a player "to get under all conditions"[10].

The three options above are the only strategic possibilities of the game and the reason why the imputation (1/3, 1/3, 1/3) does not make sense as an initial bargaining result but as a second order consideration where the players realize[11] that only in two out of the three possibilities they can obtain ½. When there is no reason to believe one coalition to be more likely to form than another one, then, each player can expect to get 2/3 * ½ = 1/3. They could agree on accepting 1/3 as the expected value of the

---
[10] See vN-M page 228.
[11] We refer to chapter … moves of the third-kind in chapter ……

game. The imputation (1/3, 1/3, 1/3) could well be considered the ideal imputation for an arbitration scenario.

At this point we make an important introductory observation: the vector $x^\circ = (x_1^\circ, x_2^\circ, x_3^\circ)^t$ taken as a vector of maximum sustainable claims under all possible conditions is actually the solution to the system $W x = w$, where

$$W = \begin{pmatrix} 1 & 1 & 0 \\ 1 & 0 & 1 \\ 0 & 1 & 1 \end{pmatrix}$$ is the characteristic matrix

of the collection $C = \{\{1, 2\}, \{1, 3\}, \{2, 3\}\}$,

and $w = \begin{pmatrix} 1 \\ 1 \\ 1 \end{pmatrix}$ is the vector of characteristic

function values corresponding to the two-players coalitions.

The *extended imputation* x° of potential claims and the supporting *covering-collection* of two-players subsets of N ={1,2,3} that support them conform the pair (x°, C) that will be acknowledge as the system that defines the *strategic equilibrium* for the game.

The strategic-equilibrium, in the 3-person zero-sum game happens to be the nondiscriminatory solution x° =(½, ½, ½), C = {{1,2}, {1,3}, {2,3}} may be better understood as the system of interrelated and mutually exclusive potential outcomes given bellow (½, ½, 0) if {1, 2} {3} occurs

$\qquad$ (½, 0, ½)   if   {1, 3} {2} occurs

$\qquad$ (0, ½, ½)   if   {2, 3} {1} occurs

Any player can test the stability of the above conditional solution system simply by requesting his partner, a utility transfer $0 < \varepsilon < ½$. Say player 1 request a utility transfer from player 2 so that the departing point is

$\qquad$ (½, ½, 0)   if   {1, 2} forms

The request

$\qquad$ (½ +ε, ½ -ε, 0)   if   {1, 2} forms

Player 2 would simply point out to player 1 that such a request is not sound, for if he thinks he may obtain the amount ½ +ε elsewhere, it would have to be using his only bargaining alternative with player 3 and the result would be

$$(\tfrac{1}{2}+\varepsilon,\ 0,\ \tfrac{1}{2}-\varepsilon) \quad \text{if} \quad \{1,3\} \text{ forms}$$

But such result would be unstable for you player 1 because I, player 2, can offer player 3 the amount $\tfrac{1}{2} > \tfrac{1}{2}-\varepsilon$ and retain the actual amount of $\tfrac{1}{2}$ in the potential outcome

$$(0,\ \tfrac{1}{2},\ \tfrac{1}{2}) \quad \text{if} \quad \{2,3\} \text{ forms}$$

We will see later that, covering collections that support an extended imputation conform a strategic equilibrium if the structure of the collection does not admit utility transfers case in which the collection must be linearly balanced. The following graph for the 3-person zero-sum game makes explicit the comments above

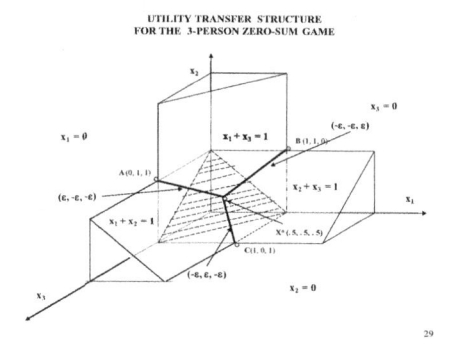

Figure 1.8

In this case, as in vN-M composition and decomposition of imputations, our extended imputation x° corresponds to the " ...operation of "viewing as one" ..." [12], in our case, three separate and mutually exclusive occurrences.

So far, to introduce the strategic-equilibrium of the game we have use vN-M observations used to obtained the "objective" non-discriminatory solution of the 3-person zero-sum game. However, vN-M concept of *claims that players can reasonably expect to get under all conditions*[13], though used not in an explicit manner in the computation of the solutions to the 4-person zero sum game and in the approach for the determination of all solutions to the general-sum n-person game, was never, explicitly defined, mathematically characterized, nor explicitly declared as a fundamental equilibrium of the game.

---

[12] See vN-M page. 360
[13] These are computed for the simple majority game when coalitions have different strength vN-M pg. 230-233 ,for the 3-person zero sum game in pp. 260-261, and pp. 297-298 for the 4-person zero-sum game.

1.3 Genesis of a strategic-equilibrium for the game

We have mentioned von Neumann and Morgenstern basic argument on *claims* that can be reasonably maintained under all conditions. Further, the remarks made about the "objective" non-discriminatory solution clearly uncover an undisputable strategic equilibrium for the game. It exhibits a stability that can only be conceived in terms of interrelated mutually exclusive occurrences that are viewed as an integrated whole. The fact that this particular equilibrium happened to possess the stability required for a vN-M solution in terms of domination was perhaps the critical obstacle to the formulation of an independent mathematical characterization of such "conspicuous" systemic attractor clearly emerging from the exercise of reasonable bargaining alternatives in a cooperative game.

Clearly, the discriminatory solutions possess vN-M stability characteristics related to domination but don't have all the characteristics that define the objective solution.

Vickrey's *strong solutions* may be considered the first attempt to characterize the equilibrium of the "symmetric", "objective" non-discriminatory solution but mainly in the line of narrowing down the span of "enormous" possibilities exhibited by vN-M solutions. His argument was based on an unavoidable cycle that would lead to the punishment necessarily suffered by ambitious defectors of the strong solution. However, the characterization of those properties, though successful for the 3-person zero-sum game, didn't go too far[14]. Milnor's *reasonable outcomes* was another attempt to capture stronger equilibrium conditions than those of vN-M stable sets, Maschler objections and counter-objections argument for *bargaining sets* and the related excess equilibrium argument in the *kernel,* may be thought as other attempts to capture the unquestionable equilibrium of the symmetric solution not present in the discriminatory solutions. The last three mentioned approaches have in common the fact that they center their analysis on payoffs associated with partitions. After all, a partition of the set of players N, which may be the proper subset N itself, and

---

[14] See luce and raiffa

an associated payoff to the players is the inevitable final outcome in any cooperative game. In relation to the 3-person zero sum game they manage to exhibit all the outcomes included in the vN-M symmetric solution. Nonetheless these solutions appear as fragmented items voided of the systemic interrelations explicitly emphasized in vN-M analysis where the elements in the solution set cannot be considered solutions by themselves but as an interrelated whole.

It appears that the equilibrium that characterized the symmetric solutions is not a property that can easily be found in any particular outcome of a given game. Intuitively, we anticipate the complexity of looking for the causes of something by focusing on the characteristics its results. Instead of focusing on partitions as the central unit of analysis as with the *individually rational payoff configurations* of the kernel and the bargaining sets, we focus on covering collections or simply, collections of subsets of N that are a cover of N. Partitions are a particular case of covers of N. Precisely partitions constitute the case of collection of sets voided of strategic interactive significance to define equilibrium for they contain no bargaining alternatives for the players.

With the hope of capturing the essential characteristics of the symmetric solution the author presented in his doctoral dissertation the concept of structural and strategic equilibriums and characterized them in term of *linearly balanced collections*. The results rely on a heuristic analysis of *utility transfers* on *attainable extended-imputations*. These approaches have been formalized through rigorous mathematical characterizations[15] that are beyond the scope of this presentation which is based mainly on heuristic arguments developed through a constructive approach. We will introduce the strategic equilibrium using the basic arguments originally developed in von Neumann and Morgenstern namely, when they identify the maximum "claims attainable under all possible conditions" by the players.

The approach followed here is consistent with von Neumann and Morgenstern recommendation in solving

---

[15] See chapter on Utility Transfer Analysis

the general sum n-person cooperatgame game with transferable utility of looking first for the objective solution:

The fundamental strategic equilibrium introduced here, is one about possibilities. It is the result of focusing mainly on emergent properties and stable structures of the bargaining process, which may take place in a cooperative game. It characterizes stability on structures of contingent realizable payoffs, given by claims that may be attainable but conditioned to the actual formation, of possible coalitions that may support them. This equilibrium may well be considered a generalization of the one in (vN-M) von Neumann and Morgenstern (1944) p.261, found for the three-person zero-sum game and expressed in the symmetric "objective" non-discriminatory solution. As in the analysis of the above-mentioned game, the heuristic process by which the fundamental equilibrium is identified, for all cooperative games with transferable utility, is based on establishing whether or not players' claims, in excess of certain payoff level, can be reasonably maintained under all possible conditions.

A retrospective consideration of von Neumann and Morgenstern's theoretical development, allows us to think of the symmetric solution system (equilibrium), as one coalitionally rational *extended imputation* realizable by parts, not as a whole. This type of payoff vector is the composition of the three mutually exclusive and collectively exhaustive potential occurrences, represented by three payoff vectors of maximum sustainable claims (also extended imputations), each one having as effective set one of the two-person coalitions that may form. Hence, the "objective" symmetric solution to the three-person zero-sum game, as a system, may be described: either as a pair [$x°$, C] where $x° = (½, ½, ½)^t$ is an extended imputation and $C = \{1,2\}, \{1,3\},\{2,3\}\}$ is a cover of N that supports $x°$; or equivalently, as a *conditional matrix* describing an interrelated system of contingent payoffs where each row gives the amounts that players may receive if the corresponding coalition forms:

(1)
$$\begin{bmatrix} (1/2 & 1/2 & 0) & if & \{1,2\}\, forms \\ (1/2 & 0 & 1/2) & if & \{1,3\}\, forms \\ (0 & 1/2 & 1/2) & if & \{2,3\}\, forms \end{bmatrix}.$$

To capture the equivalent systemic stability imbedded in the general n-person cooperative game with transferable utility, we will depart considerably from traditional approaches. Instead of working with imputations (individually rational payoffs constrained by the grand coalition) as in von Neumann and Morgenstern's solutions(1944) and in the core of Gilles (1959), or focusing in rational payoff configurations (partitions with corresponding payoffs) as in Aumann and Maschler's (1964) in the theory of the Bargaining Sets and in Davis and Maschler's (1965) theory of the Kernel, we use for payoff bargaining analysis in coalition formation, *extended imputations* supported by *cover collections* not necessarily partitions. This significant departure[16] will allow us to better understand the systemic interrelation and role that bargaining alternatives may play, in defining the coalitions likely to form in a stage previous to the possible formation of the grand coalition. It will also help

---

[16] The need to break away with the "feasibility trap", i.e . to work exclusively with payoff vectors that add up to the value of the game for the grand coalition, has been a recurrent theme in the theory of coalition formation. See for example,
vN-M (1944), Cross (1963), Aumman and Maschler(1964), Turbay (1977), Selten (1981), Albers(1987) , Bennett(1983), Moldavanu and Winter (1994), among many others, and explicitly manifested against it in Bejan and Gómez (2012).

us to determine the manner in which the corresponding proceeds, of the coalitions that may form, are to be divided among its members. One would expect rational players to have the means of knowing ¿how well they may do in the game? before considering joining the grand coalition. The fundamental equilibrium provides a definitive answer to this question. In so doing, it also gives a partial answer to the questions that were once raised by Michael Maschler when he asked which coalitions are likely to form if the grand coalition doesn't form; and how the proceeds of a coalition that may form are to be divided among its members. A complete answer to these questions requires the resolution of subsequent bargaining stages.

The fundamental strategic equilibrium is a well-defined mathematical object shown to exist for every e-essential cooperative game[17]. It is given either as (1) a bargaining pair $[x^\circ, C]$ where x° is an un-dominated extended imputation with p-balanced support C or as (2) a contingent system of C-feasible vectors described by the rows of a kxn matrix $X^\circ = (x^\circ_{ij})$ with the same zero pattern

---

[17] It includes vN-M essential supper-additive games and Aumman and Maschler non-supper-additive as special cases.

of the characteristic matrix W of the p-balanced collection C that supports $x°$ where $x°_{ij} = x_j°$ if $w_{ij} = 1$ and $x°_{ij} = 0$ if $w_{ij} = 0$. In both descriptions, the fundamental strategic equilibrium is to be taken as a conditional system of C-feasible (realizable) payoffs imbedded into a single coalitionally rational extended imputation $x°$ attainable through the cover C of N.

The existence of a detached extended imputation x^* that satisfy all rationality conditions including a relaxed form of group rationality $(x^*(N) \geq v(N))$, was first established by von Neumann and Morgenstern (1944) p.370. In fact, they proved that every game Γ determines a number $|\Gamma_2|$ with the property, among several, that "a detached extended imputation with excess e exists if and only if $e \geq |\Gamma_2|$" (here $e(x) = v(N) - x(N)$) and $|\Gamma_2| = e^*$ with $e(x^*) = \min e(x)$ over all x detached. Clearly x^* that minimizes[18] $e(x)$ over x detached also minimizes the sum $x(N) = x_1 + \ldots + x_n$ over $\{x | x(S) \geq v(S)$ for all $S \subseteq N\}$. It becomes clear today, that this result implicitly gives the necessary and sufficient conditions for the existence of the

---

[18] The existence of the detached extended imputation x^* that minimizes $e(x)$ was proven using continuity not linear programming

core of the game Γ. This is so, because if there is a detached extended imputation which is also an imputation then such imputation must be in the core of Γ. It also becomes clear that the results of Bondareva (1963) and Shapley (1967) on the existence of the core may be viewed as the equivalent dual formulation of the conditions for the existence of a detached extended imputation with $e = 0$ of vN-M excess minimization problem which is a basic linear programming problem that has been often associated with the general sum n-person cooperative game in characteristic function form.

The stability properties of extended imputations which satisfy individual and coalitional rationality conditions were independently studied by Cross (?), Albers (?) and Turbay (?). Cross' *stable demand vectors* and Turbay's *strategically stable extended imputations* both correspond to extended imputations that may be attainable through a cover collection of subsets of N. Bennett (1984) develops the *aspiration approach* with similar and equivalent concepts and extended existing solution concepts at the time, such as vN-M stable sets and the Aumman and Maschlesr's bargaining set to the aspirations domain. In fact, aspirations may be defined using vN-M extended imputations and Turbay's cover collections as detached extended imputations with cover support. Similarly to Harsanyi's reinterpretation (?) of vN-M solution concept in terms of equilibrium points, Selten (1981), referring to the mentioned above developments of Cross, Albers and Turbay, gives a reinterpretation of the stable demand vectors for the cooperative game in characteristic function form and shows for an associated sequential proposal non-cooperative bargaining game that the sub-game perfect stationary equilibriums of the latter and the stable demand vectors of the former imply each other. It is interesting to note that the first stage solution proposed for the general-sum cooperative game in Turbay (1977) becomes *de facto*, by Selten's theorem a subgame perfect equilibrium of the associated sequential proposal bargaining game. Harsanyi's and Selten's pioneering approaches may be considered together with Nash's program the initiators of the non-cooperative approach[19] to under-standing cooperative game theory. An extensive line of research that

---

[19] The Nash program, being perhaps, the most ambitious proposal in this direction.

studies endogenous coalition formation and axiomatic approaches that includes Bennett critique and refinement of Selten's original results Mold ... ---- ...... Wintwr, Gom.mz. Though the non-cooperative approach to cooperative games dominates today's research landscape in coalition formation and enhances the understanding of cooperative game theory. It is among the main purposes of this paper to emphasize the strategic nature of cooperative games as initially presented by vN-M and to retake the concepts of domination and stable sets in an extended sense, as the key to develop a solid theory rational behavior in coalition formation. We want also to emphasized de systemic nature of the identified fundamental equilibrium and the fact that it may be considered a generalization of vN-M objective solution to the 3-person zero-sum game and that to obtain it should be, as suggested by vN-M (1944/1957) the first step toward the solution of the n-person general sum cooperative game.

Here, for the mathematical characterization of the fundamental equilibrium of the general n-person cooperative game, we will follow a constructive approach within a systems perspective. Based on utility transfers considerations, an emergent bargaining stability is observed and characterized by means of balanced collections and theorems on the alternative for matrices. A heuristic procedure to find the fundamental equilibrium by means of a player's auction is given. Further stability properties are given by theorems relating balanced sets to the concepts of *strategic independence* and bargaining alternatives. Utility transfers here are intended to capture the required compensations among players to keep coalitions from dissolving when confronted with bargaining alternatives presented as optional payoff agreements to be subscribed by the players.

An equivalent representation of the fundamental equilibrium is shown to be obtainable from the solution of a well-known dual pair of linear programming problems, associated with the general cooperative game with transferable utility. Such representation of the equilibrium was originally introduced in Turbay (1977). There, the set of all strategic equilibriums with realizable payoffs satisfying individual and coalitional rationality was identified as the first stage solution to the general cooperative game. The above-mentioned dual pair of linear programming problems are often found in relation to the theorems on the existence of the core Bondareva (1963), Shapley (1965), Charnes and Kortanek's (1966) and Scarf (1967); and also, in relation to the *aspirations* approach to solution concepts in Bennett (1983), Albers (1987), and Bennett, Maschler and Zame's (1994). Thus, we are able to take advantage of all the power of linear programming, in establishing important properties of the fundamental equilibrium, such as those of existence and optimality (overall strategic equilibriums). Other properties such as invariance under strategic equivalence are also proven.

In addition to its possible interpretation as a solution concept, we will emphasize here a bargaining attractor characteristic of the fundamental equilibrium, for we believe it should be considered as an initial stage solution or bifurcation point where from a system of multiple alternative equilibrium realities may unfold as the solution to the n-person cooperative game. We believe, this re-introduction of the strategic equilibrium concept, opens many new perspectives, providing a highly positive outlook for the theory of cooperative games with transferable utility and its applications.

The fundamental strategic equilibrium as a solution concept differs radically from previous solution concepts in that it is not simply a payoff in a given domain, but, instead, it consists of a system of coalitionally feasible payoff vectors, with corresponding supporting coalitions whose stability, consistent with von Neumann and Morgenstern expectations for a solution[20], is a property of the system as a whole and not of the payoffs of which it is composed. All coalitionally feasible payoffs are composed into a single extended imputation. Say, as an omniscient view of multiple possible realities. Different contingent payoffs are assembled together and seen as one point. However, each point is tied to an interrelated internally stable coalitional system of possible occurrences, generally of mutually exclusive nature.

If one of the coalitions associated with a fundamental equilibrium forms, then the players may have to decide whether to: (1) take the prescribed equilibrium payoffs as final payoff in the game, or (2) accept these payoffs with the binding agreement not to dissolve by

---

[20] See vN-M p.36

external offers (form a syndicate); and then to bargain as a unit against the players in the complementary coalition for their marginal contribution[21] to the coalition of all players. Those players in the complement to N of a formed coalition, may have to play the corresponding restriction of the original game. The accepted original equilibrium payoffs, by the players of a formed syndicate, become *de facto* disagreement dividends. Clearly this analysis has to be carried out for all coalitions that are likely to form. Thus, the identified syndicate bargaining process represents a second stage of the game that needs to be resolved in the path towards obtaining the complete solution of the general n-person cooperative game with transferable utility. Here we will concentrate, exclusively, in the first stage solution of the game, namely the one where the fundamental equilibrium is obtained.

We arrive to two major conclusions in our equilibrium analysis: (1) the fundamental strategic equilibrium(s) imbedded in every cooperative game, as

---

[21] The marginal contribution and the bargaining process departing from the un-dominated equilibrium payoffs here is equivalent to the one described in von Neumann and Morgenstern (1944) p.417-418, describing the solution curves as a result of players in a formed coalition bargaining for the positive excess with the excluded player.

mentioned above, may be viewed not necessarily as a solution concept but as a fundamental bargaining attractor from which all possible solutions to the general cooperative game may be better understood and explained. And (2) we have acquired the perspective of a proper understanding of the general n-person cooperative game and its possible solutions requires the recognition of qualitatively different bargaining stages and the identification of their corresponding outcomes. These stages and outcomes emerge, unfold, and are seen to be interconnected in sequential form, whenever the n-person cooperative game is approached from a constructive rather than axiomatic or deterministic point of view.

We hope these preliminary results to be indicators consistent with Harsanyi's (1961) views and Bradenburger's (2007) considerations on the relevance of cooperative games (interview with Hendricks (2007)). And also, in agreement with Aumann's perception of the revival and importance of these games, in the general context of game theory (interview with van Damme (1998)). We believe the systems approach to stability presented here possess a higher explicative capacity than previous approaches. We expect these and other new results around the fundamental equilibrium soon to become a major contributing factor to the mentioned "revival" trend, the basis for a unified theory of solutions for the n-person cooperative game, and a major tool in applications to understanding and modeling economic and social interactive behavior.

2. Auction Equilibrium in Coalition Formation

Let us consider the 0-normalized 3-person cooperative game $\Gamma = (N, v)$ with characteristic function given in parametric form by:

$v(\emptyset) = v_0 = 0, \quad v(\{1\}) = 0, \quad v(\{2\}) = 0, \, v(\{3\}) = 0,$

$v(\{1, 2\}) = v_{12}, \, v(\{1, 3\}) = v_{13}, \, v(\{2, 3\}) = v_{23} \text{ and } v(\{1, 2, 3\}) = v_{123}$

Suppose a player is considering proposing a competitive bidding for his cooperation in forming a coalition with the highest bidder among potential partners. As in a Japanese Auction, let us allow for the selling player to call out an ascending value β. All interested players are in until dropping out at a certain value level. The player remaining when the last to second drops out is the one to be the selected partner. Here we want to find the maximum call value β^* at which the bidding players drop out for a better bargaining alternative.

Let us denote the three players with the letters i, j and k and x_i, x_j and x_k respectively, the variable amounts they will end up receiving at any given moment during the bidding process. Let's assume that player i is the auctioneer. Then, at a given call value β, in the two possible coalitions that may form, the amounts the players may obtain are given in the following table:

	X_i	X_j	X_k	
β (2.1)	$(v_{ij} - \beta)$	0		if $\{i,j\}$ forms
β (2.2)	0		$(v_{ik} - \beta)$	if $\{i,k\}$ forms

The amounts received by the players must be non-negative satisfying individual rationality conditions $x_i \geq 0$, $x_j \geq 0$ and $x_k \geq 0$. These become

$$\beta \geq 0, \quad (v_{ij} - \beta) \geq 0 \quad \text{and} \quad (v_{ik} - \beta) \geq 0 \quad (2.3)$$

Clearly, players j and k are willing to join player i as long as the sum of the amounts they obtain is greater than or equal to but not less than the amount they can obtain by joining together in coalition $\{j, k\}$. That is, whenever the coalitional rationality conditions $x_j + x_j \geq v_{jk}$ are satisfied for players i, j, and k.

$$(v_{ij} - \beta) + (v_{ik} - \beta) \geq v_{jk}; \quad i,j,k = 1,2,3\,;\, i \neq j \neq k$$

Or equivalently

$$\beta \le \frac{(v_{ij} + v_{ik} - v_{jk})}{2} \quad i,j,k = 1,2,3 \ ; \ i \ne j \ne k \quad (2.4)$$

We can readily check that both rationality conditions (2.3) and (2.4) are satisfied if and only if the conditions

$$0 \le \beta_i \le min\left[\frac{(v_{ij} + v_{ik} - v_{jk})}{2}, v_{ij}, v_{ik}\right] \quad i,j,k = 1,2,3; \ i \ne j \ne k \quad (2.5)$$

are satisfied for each of the player's bidding call value $\beta_i, i = 1, 2, 3$.

Without loss of generality, we may assume $v_{12} \ge v_{13} \ge v_{23}$. And let β_i^* be the maximum bidding call value for player i before drop-out, i= 1, 2, 3.

2.1 Bidding for player 1's cooperation

The rationality conditions summarized by (2.5) in the bidding process for the cooperation of player 1 imply

$$0 \le \beta_1^* = min\left[\frac{(v_{12} + v_{13} - v_{23})}{2}, v_{13}\right] \quad (2.6)$$

we can verify that the maximum bid for player 1 is

$$\beta_1^* = \begin{cases} \dfrac{(v_{12} + v_{13} - v_{23})}{2} & \text{if and only if} \quad v_{12} \leq v_{13} + v_{23} \quad (2.7) \\ v_{13} & \text{if and only if} \quad v_{12} > v_{13} + v_{23} \quad (2.8) \end{cases}$$

The bidding value β_1^* is clearly the maximum amount that either player 2 or player 3 would be willing to give to play 1 in forming a coalition. If player 1 receives $x_1 = \beta_1^*$ then players j and k will receive $x_j = (v_{12} - \beta_1^*)$ and $x_k = (v_{13} - \beta_1^*)$ respectively provided the corresponding coalition forms. Note that at bidding value β_1^*, player 1 is indifferent in joining either one of the coalitions $\{1, 2\}$ or $\{1, 3\}$.

A 3-person cooperative game is said to be a *triangular game* if and only if

$$v_{12} \leq v_{13} + v_{23} \qquad (2.9)$$

Otherwise, whenever

$$v_{12} > v_{13} + v_{23} \qquad (2.10)$$

holds, it is said to be **non-triangular**. Thus, we have the following two cases:

2.1.1 Bidding for player 1 in a triangular game

$(v_{12} \leq v_{13} + v_{23})$. Here, the amount that player 1 can obtain in (2.1) and in (2.2) at the dropping-out level is $x_1^o = \beta_1^*$. The complementary values received at this level by players 2 and 3 are $x_2^o = (v_{ij} - \beta_1^*)$ and $x_3^o = (v_{ik} - \beta_1^*)$ respectively. Since the game is assumed to be triangular, then (2.7) holds and $\beta_1^* = \frac{(v_{12}+v_{13}-v_{23})}{2}$. Thus, we obtain:

$$x_1^o = \frac{(v_{12} + v_{13} - v_{23})}{2}$$

$$x_2^o = \frac{(v_{12} - v_{13} + v_{23})}{2} \qquad (2.11)$$

$$x_3^o = \frac{(-v_{12} + v_{13} + v_{23})}{2}$$

Considering the contingencies (2.1) and (2.2) at the drop-out point, with i=1, j=2 and k=3 namely, coalitions {1, 2} and {1, 3} with corresponding associated payoffs together with the possible formation of the alternative coalition {2, 3} and its corresponding payoffs, we obtain the following conditional system which show us what the players may receive in each possible occurrence:

x_i	x_j	x_k	Coalition Structure
$(x_1^o$	x_2^o	$0)$ if $\{1,2\}\{3\}$	forms
$(x_1^o$	0	$x_3^o)$ if $\{1,3\}\{2\}$	forms (2.12)
$(0$	x_2^o	$x_3^o)$ if $\{2,3\}\{1\}$	forms

Here, we may verify that

$$x_1^o + x_2^o = v_{12}, \quad x_1^o + x_3^o = v_{13} \text{ and } x_2^o + x_3^o = v_{23} \quad (2.13)$$

The specific values for $x_l^o, l = 1,2,3$ are obtained from the equations (2.11) above.

We may assert that coalitions in $\mathbb{C} = \{C_1, C_2, C_3\}$, $C_1 = \{1,2\}, C_2 = \{1,3\}, C_3 = \{2,3\}$ with corresponding payoffs summarized in the extended imputation $x^o = (x_1^o, x_2^o, x_k^o)$ are the coalitions likely to form. But only one of them may form and he proceeds of the coalition that forms are divided accordingly.

2.1.2. Bidding for player1 in a non-triangular game

$(v_{12} > v_{13} + v_{23})$.

Similar considerations as the above ones but for a non-triangular game give us $x_1^o = \beta_1^*$,

$x_2^o = (v_{12} - \beta_1^*)$ and $x_3^o = (v_{13} - \beta_1^*)$.

Since $\beta_1^* = v_{13}$, we obtain:

$$x_1^o = v_{13}$$
$$x_2^o = v_{12} - v_{13} \qquad (2.14)$$
$$x_3^o = 0$$

However, at the drop-out point for player 3 the alternative coalition {2, 3} with $x_3^o = 0$ gives another possible value for player 2, namely $x_2^o = v_{23}$. This emerging possibility for player 2 can be thought of as player 2's drop-out value level at which he would have a bargaining alternative, if the biding level is allowed to increase beyond the second to last drop-out. Such increase would require a modification of our initially proposed Japanese auction. To keep the Japanese bid structure unmodified, we could assume the presence of a stooge or dummy bidder that remains during auction until last player withdraws from the auction. Thus values v_{13} and v_{23} may be taken as disagreement values in a negotiation between players 1 and 2 for the remaining value $\Delta = v_{12} - (v_{13} + v_{23})$ which they can create.

Consequently, we may expect coalition {1, 2} to be the one most likely to form with resulting payoffs to players 1 , 2 player 3 given by :

$$x_1^o = v_{13} + \alpha \Delta$$
$$x_2^o = v_{23} + (1-\alpha)\Delta \qquad (2.15)$$
$$x_3^o = 0 , \quad 0 \leq \alpha \leq 1$$

Clearly the values of x_1^o and x_2^o in (2.15) equal those in (2.14) for $\alpha = 1$.

Remark 2.1 The indeterminacy that remains is the one characteristic of the pure bargaining 2-person game cooperative game. ***The Nash bargaining solution*** is one of the most common ways of reducing uncertainty in this type of scenarios. Such indeterminacy problem will be addressed in later chapters where we will introduce and propose a *most preferent solution* concept for pure bargaining scenarios.

2.2 Bidding for player 2's cooperation

The auction process described through conditions and equations (2.1} to (2.5) applies equally for all three players Thus for player 2 we must have i= 2 in (2.5) and

$$0 \le \beta_2^* = \min\left[\frac{(v_{12} - v_{13} + v_{23})}{2}, v_{23}\right] \qquad (2.16)$$

The maximum bid for player 2 is

$$\beta_2^* = \begin{cases} \dfrac{(v_{12} - v_{13} + v_{23})}{2} & \text{if and only if} \quad v_{13} \le v_{12} + v_{23} \quad (2.17) \\ v_{23} & \text{if and only if} \quad v_{13} > v_{12} + v_{23} \quad (2.18) \end{cases}$$

2.2.1 Bidding for player 2 in a triangular game ($v_{12} \le v_{13} + v_{23}$).

Here, the amount that player 2 can obtain in (2.1) and in (2.2) at the dropping-out level is $x_2^o = \beta_2^*$. The complementary values received at this level by players 2 and 3 are $x_2^o = (v_{12} - \beta_2^*)$ and $x_3^o = (v_{23} - \beta_2^*)$ respectively. Since the game is assumed to be triangular, then (2.17) holds and $\beta_2^* = \frac{(v_{12} - v_{13} + v_{23})}{2}$. Thus, we obtain:

$$x_1^o = \frac{(v_{12} + v_{13} - v_{23})}{2}$$

$$x_2^o = \frac{(v_{12} - v_{13} + v_{23})}{2} \qquad (2.19)$$

$$x_3^o = \frac{(-v_{12} + v_{13} + v_{23})}{2}$$

Considering the contingencies (2.1) and (2.2) at the drop-out point with i=2, j=1, k=3 namely, coalitions {1, 2} and {2, 3} with corresponding associated payoffs together with the possible formation of the alternative coalition {1, 3}and its corresponding payoffs, we obtain the same conditional system as the one obtains in the auction for player 1 in (2.1.1)which show us what the players may receive in each possible occurrence:

$$
\begin{array}{cccll}
X_i & X_j & X_k & & \\
\text{Coalition Structure} & & & & \\
\hline
(x_1^o & x_2^o & 0\ ) & if & \{1,2\}\ \{3\}\ forms \\
(x_1^o & 0 & x_3^o\ ) & if & \{1,3\}\ \{2\}\ forms \quad (2.20) \\
(0 & x_2^o & x_3^o\ ) & if & \{2,3\}\ \{1\}\ forms \\
\end{array}
$$

Here, we may verify again that

$$x_1^o + x_2^o = v_{12},\ x_1^o + x_3^o = v_{13}\ and\ x_2^o + x_3^o = v_{23} \quad (2.21)$$

The specific values for $x_j^o, j = 1,2,3$ are obtained from the equations (2.19) above.

We may assert that coalitions in $\mathbb{C} = \{C_1, C_2, C_3\}$, $C_1 = \{1,2\}$, $C_2 = \{1,3\}$, $C_3 = \{2,3\}$ with corresponding payoffs summarized in the extended imputation $x^o = (x_1^o, x_2^o, x_k^o)$ are the coalitions likely to form. But only one of them may form and the proceeds of the coalition that forms are divided accordingly.

2.2.2. Bidding for player2 in a non-triangular game

($v_{12} > v_{13} + v_{23}$). Similar considerations as the above ones but for a non-triangular game give us

$$x_1^o = (v_{12} - \beta_2^*), \quad x_2^o = \beta_2^* \text{ and } x_3^o = (v_{23} - \beta_2^*).$$

Since $\beta_2^* = v_{23}$, we obtain:

$$x_1^o = v_{12} - v_{23}$$
$$x_2^o = v_{23} \qquad (2.22)$$
$$x_3^o = 0$$

As in the auction for player 1 above, at the drop-out point for player 3 the alternative coalition {1, 3} with $x_3^o = 0$ gives another possible value for player 1, namely $x_1^o = v_{13}$. This emerging possibility for player 1 can be thought of as player 1's drop-out value level at which he would have a bargaining alternative, if the biding level is allowed to increase beyond the second to last drop-out. Such increase would require a modification of our initially proposed Japanese auction. To keep the Japanese bid structure unmodified, we could assume the presence of a stooge or dummy bidder that remains during auction until last player withdraws from the auction. Thus values v_{12} and v_{23} may be taken as disagreement values in a negotiation between players 1 and 2 for the remaining value $\Delta = v_{12} - (v_{13} + v_{23})$ which they can create. Consequently, we may expect coalition {1, 2} to be the one most likely to form with resulting payoffs to players 1, 2 player 3 given by :

$$x_1^o = v_{13} + \alpha\Delta$$
$$x_2^o = v_{23} + (1-\alpha)\Delta \qquad (2.23)$$
$$x_3^o = 0 , \quad 0 \leq \alpha \leq 1$$

As in the auction for player 1, the values of x_1^o and x_2^o in (2.23) equal those in (2.22) for $\alpha = 1$, and Remark 2.1 applies likewise.

2.3 Bidding for player 3's cooperation

As in the preceding two auctions, the auction process described through conditions and equations (2.1} to (2.5) which applies equally for all three players, we note that $\frac{(-v_{12}+v_{13}+v_{23})}{2} \leq v_{23}$ always, because $v_{12} \geq v_{13} \geq v_{23}$. And hence, in this case for i= 3 we get in (2.5).

$$\beta_3^* = min\left[0, \frac{(-v_{12} + v_{13} + v_{23})}{2}\right] \qquad (2.24)$$

Thus, the maximum bid for player 3 is

$$\beta_3^* = \begin{cases} \frac{(-v_{12} + v_{13} + v_{23})}{2} & \text{if and only if} \quad v_{12} \leq v_{12} + v_{23} \quad (2.25) \\ 0 & \text{if and only if} \quad v_{13} > v_{12} + v_{23} \quad (2.26) \end{cases}$$

2.3.1 Bidding for player 3 in a triangular game ($v_{12} \leq v_{13} + v_{23}$).

Here, the amount that player 3 can obtain in (2.1) and in (2.2) at the dropping-out level is $x_3^o = \beta_3^*$. The complementary values received at this level by players 1 and 2 are $x_1^o = (v_{13} - \beta_3^*)$ and $x_2^o = (v_{23} - \beta_3^*)$ respectively. Since the game is assumed to be triangular, then (2.25) holds and $\beta_3^* = \frac{(-v_{12}+v_{13}+v_{23})}{2}$. Thus, we obtain:

$$\begin{aligned} x_1^o &= \frac{(v_{12} + v_{13} - v_{23})}{2} \\ x_2^o &= \frac{(v_{12} - v_{13} + v_{23})}{2} \qquad (2.27) \\ x_3^o &= \frac{(-v_{12} + v_{13} + v_{23})}{2} \end{aligned}$$

Considering the contingencies (2.1) and (2.2) at the drop-out point with i=3, j=1, k=2 namely, coalitions {1, 3} and {2, 3} with corresponding associated payoffs together with the possible formation of the alternative coalition {1, 2} and its corresponding payoffs, we obtain the same conditional system as the one obtains in the auctions for players 1 and 2 in (2.1.1) and (2.2.1) which show us what the players may receive in each possible occurrence:

$$
\begin{array}{ccc}
X_i & X_j & X_k
\end{array}
$$

Coalition Structure

$$
\begin{array}{llll}
(x_1^o & x_2^o & 0\,) & if \quad \{1,2\}\ \{3\}\ forms \\
(x_1^o & 0 & x_3^o\,) & if\ \{1,3\}\ \{2\}\ forms \quad (2.28) \\
(0 & x_2^o & x_3^o\,) & if\ \{2,3\}\ \{1\}\ forms
\end{array}
$$

Here again, we may verify that

$$x_1^o + x_2^o = v_{12},\ \ x_1^o + x_3^o = v_{13}\ \text{and}\ x_2^o + x_3^o = v_{23} \qquad (2.29)$$

The specific values for x_j^o, $j = 1,2,3$ are obtained from the equations (2.19) above.

We may assert that coalitions in $\mathbb{C} = \{C_1, C_2, C_3\}$, $C_1 = \{1,2\}$, $C_2 = \{1,3\}$, $C_3 = \{2,3\}$ with corresponding payoffs summarized in the extended imputation $x^o = (x_1^o, x_2^o, x_k^o)$ are the coalitions likely to form, but only one of them may form and the proceeds of the coalition that forms are divided accordingly.

2.3.2. Bidding for player3 in a non-triangular game ($v_{12} > v_{13} + v_{23}$). Here there is no value β_3^*, non-negative which allow players 1 and 2 to obtain respectively values $x_1^o = (v_{13} - \beta_3^*)$ and $x_2^o = (v_{23} - \beta_3^*)$ that satisfy the coalitional rationality condition $(v_{13} - \beta_3^*) + (v_{23} - \beta_3^*) \geq v_{12}$. The dropping-out level may be assumed to be $x_3^o = \beta_3^* = 0$. The complementary values received at this level by players 1 and 2 are $x_1^o = (v_{13} - \beta_3^*)$ and $x_2^o = (v_{23} - \beta_3^*)$ respectively. Thus, we obtain:

$$x_1^o = v_{13}$$
$$x_2^o = v_{23} \qquad (2.30)$$
$$x_3^o = 0$$

Since the game is non-triangular, then $\Delta = v_{12} - (v_{13} + v_{23}) > 0$ and players 1 and 2 may cooperate and player 3 remains by himself to obtain the amounts:

$$x_1^o = v_{13} + \alpha\Delta$$
$$x_2^o = v_{23} + (1-\alpha)\Delta \qquad (2.31)$$
$$x_3^o = 0, \quad 0 \leq \alpha \leq 1$$

2.4 Auction Consolidation Summary

To sum up, in the preceding auctions a similar procedure was carried out to bid for the cooperation of each of the players. In each of the three auctions we obtained the same payoff structure in each of the two possible types of 3-person cooperative games: triangular and non-triangular. Regardless of the player that was bided for, the values of the maximum bid for a player in one auction was equal to the complementary values obtained by the same player in the other two auctions. We obtained the same maximum bidding values and corresponding payoffs as those in (2.32). Then such values are attainable for each player in one and only one of the two alternative coalitions they may join. Here we will argue, following Example 2.1 below that these bidding values are maximum sustainable claims that the players can obtain in all possible situations. The corresponding vector of maximum claims for triangular games can be given as an extended imputation here denoted by $x^o = (x_1^o, x_2^o, x_k^o)$ where

$$x_1^o = \frac{(v_{12} + v_{13} - v_{23})}{2}$$

$$x_2^o = \frac{(v_{12} - v_{13} + v_{23})}{2} \qquad (2.32)$$

$$x_3^o = \frac{(-v_{12} + v_{13} + v_{23})}{2}$$

And for non-triangular games :

$$x_1^o = v_{13} + \alpha\Delta$$
$$x_2^o = v_{23} + (1-\alpha)\Delta \qquad (2.33)$$
$$x_3^o = 0, \ 0 \leq \alpha \leq 1$$

Note that the extended imputation x^o may be attainable and the corresponding amounts x_1^o, x_2^o and, x_3^o are realizable subject to the formation of the corresponding coalition. Whenever a two-player coalition $S \subset N$ forms, the complementary coalition N/S which consists of a singleton set will form by default. The player left alone obtains a value equal to zero.

Thus, we have as possible distributions of the characteristic function value of the coalitions, given as claims in the form of extended imputations, with corresponding coalition structure displayed in the following system of payoffs which was found to be common in all their auctions :

i) For triangular games:

$$X_i \qquad X_j \qquad X_k$$

Coalition Structure

$$\begin{pmatrix} \frac{(v_{12}+v_{13}-v_{23})}{2} & \frac{(v_{12}-v_{13}+v_{23})}{2} & 0 \end{pmatrix} \text{ if } \{1,2\}\{3\} \text{ forms}$$

$$\begin{pmatrix} \frac{(v_{12}+v_{13}-v_{23})}{2} & 0 & \frac{(-v_{12}+v_{13}+v_{23})}{2} \end{pmatrix} \text{ if } \{1,3\}\{2\} \text{ forms} \quad (2.34)$$

$$\begin{pmatrix} 0 & \frac{(-v_{12}+v_{13}+v_{23})}{2} & \frac{(-v_{12}+v_{13}+v_{23})}{2} \end{pmatrix} \text{ if } \{2,3\}\{1\} \text{ forms}$$

ii) For non-triangular games

$$X_i \qquad X_j \qquad X_k$$

Coalition Structure

$$\begin{array}{lll} (v_{13}+\alpha\Delta & v_{23}+(1-\alpha)\Delta & 0 \end{array}) \quad \text{if} \quad \{1,2\}\{3\} \quad \text{forms}$$
$$\begin{array}{lll} (v_{13} & 0 & 0 \end{array}) \text{ if } \{1,3\}\{2\} \quad \text{forms} \quad (2.35)$$
$$\begin{array}{lll} (0 & v_{23} & 0 \end{array}) \text{ if } \{2,3\}\{1\} \quad \text{forms}$$

We have assumed without loos of generality that $v_{12} \geq v_{13} \geq v_{23} \geq 0$. Then for x_1^o, x_2^o, x_3^o to be non-negative we must have

$$min(x_1^o, x_2^o, x_3^o) = x_3^o = \frac{(-v_{12}+v_{13}+v_{23})}{2} \geq 0$$

Or equivalently $\quad v_{12} \leq v_{13} + v_{23}$

It becomes evident that non-negativity of the payoffs occurs if and only if the game is triangular. In these games the value created by a coalition cannot be greater than the sum of the value of the other two coalitions.

We can make a geometric model of the triangular relations defined by inequalities (2.9) and (2.10) by introducing a graph. Each vertex is labeled with the player's number and the distance between two vertices is to be taken equal to the characteristic function value created by the corresponding players. See Figure 2.1 below.

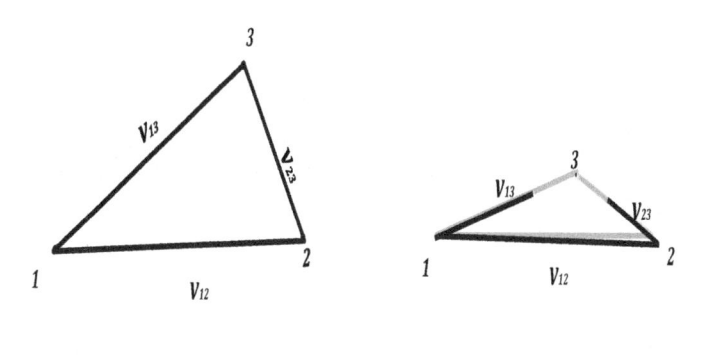

a) $v_{12} \leq v_{13} + v_{23}$
b) $v_{12} > v_{13} + v_{23}$

Figure 2.1 a) Triangular game, b) Non-Triangular game

As is to be expected, the characteristic function values of the coalitions do not make a triangle in the non-triangular case.

The contingent possible outcomes in each of the above cases can be obtained by using a ***Ruler and Compass*** construction as follows:

i) For triangular games:

Step 1. Using a ruler and a compass, construct a triangle with sides of length in proportion to the characteristic function values of the three coalitions {1, 2}, {1, 3}, and {2, 3} namely $v_{12} \geq v_{13} \geq v_{23}$ and vertices at the points 1, 2 and 3 as in Figure 2.1 above.

Step 2. Draw two circles as show in Figure 2.2 below: The first one with center at vertex 1 and radius equal to v_{13}. Denote the intersection of the circle with the side $\overline{12}$ with the letter a. Draw the second circle with center at vertex 2 and radius equal to v_{23}. Label the intersection of this second circle the side with the letter b. Use straight-edge and compass to determine the midpoint of the line segment $\overline{ba}$ and label as m_1.

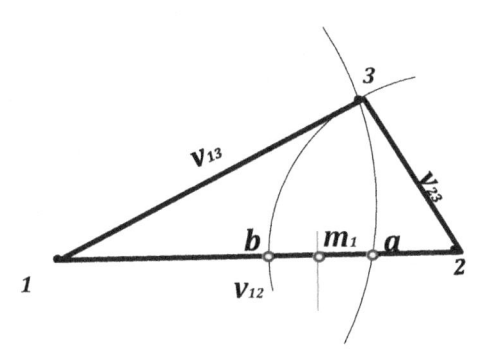

Figure 2.2 Midpoint of line segment $\overline{ba}$

Step 2. Draw a circle with center at vertex 2 and radius equal to the measure of the segment $\overline{m_1 2}$. Label the intersection of the circle with the segment $\overline{23}$ as m_2. Draw another circle with center at vertex 3 and radius equal to the measure of the segment $\overline{3m_2}$. Label the intersection of the circle with the line segment $\overline{13}$

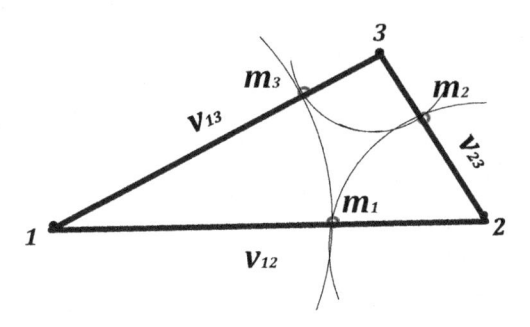

Figure 2.3 Derived points m_2 and m_3

At this point, we observe that the points m_1, m_2 and m_3 obtained by the ruler and compass construction above, divide the sides of the triangle in segments that give us the maximum sustainable claims, or equivalently the maximum bids for the players in the auction procedure introduced to explain the process of coalition formation. That is, if

$$\mu(\overline{12}) = v_{12}, \mu(\overline{13}) = v_{13}, (\overline{13}) = v_{13} = \mu(\overline{1a}), \mu(\overline{23}) = v_{23} = \mu(\overline{2b}),$$
(2.36)

Proposition 2.1 For any 0-normalized triangular game Γ= (N, v) the following always holds true

$$x_1^o = \mu(\overline{1m_1}) = \mu(\overline{1m_3}),$$
$$x_2^o = \mu(\overline{2m_2}) = \mu(\overline{2m_1}) \quad (2.37)$$
$$x_3^o = \mu(\overline{3m_3}) = \mu(\overline{3m_2})$$

Proof: By construction the point m_1 is obtained as the midpoint between the points a and b so $\mu(\overline{1m_1}) = \mu(\overline{1b}) + \frac{\mu(\overline{1a})-\mu(\overline{1b})}{2} = \frac{\mu(\overline{1a})+\mu(\overline{1b})}{2}$. Since $\mu(\overline{1a}) = v_{13}$ and $= v_{12}-v_{23}$, it follows that $\mu(\overline{1m_1}) = \frac{v_{12}+v_{13}-v_{23}}{2} = x_1^o$. Now, $\mu(\overline{2m_2}) = \mu(\overline{2m_1}) = v_{12} - \mu(\overline{1m_1}) = \frac{v_{12}-v_{13}+v_{23}}{2} = x_2^o$. And finally $\mu(\overline{3m_3}) = \mu(\overline{3m_2}) = v_{23} - \mu(\overline{2m_2}) = \frac{-v_{12}+v_{13}+v_{23}}{2} = x_3^o.$ //

The results in (2.37) are summarized in Figure 2.4 below.

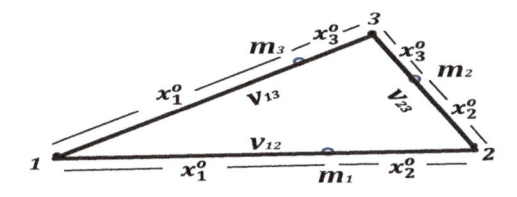

Figure 2.4 Construction of maximum bids for a triangular game

For non-triangular games, the coalition most likely to form is coalition {1, 2}. Since v_{13} and v_{23} are the maximum bids for players 1 and 2 respectively and in both cases the drop out value for player 3 is $x_3^o = 0$. Then it is clear that if coalition {1, 2} forms, these maximum bids become *de facto* disagreement payoffs and we may reasonably expect that any possible agreement between players 1 and 2 is a point in the segment $\overline{ab}$ not necessarily m_1.

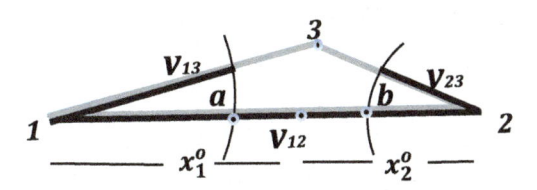

Figure 2.5 Construction of maximum bids for a non-triangular game

Since the measure of the segment $\overline{ab}$ is given by $\Delta = \mu(\overline{1b}) - \mu(\overline{1a}) = v_{12} - (v_{13} + v_{23}) > 0$ and $x_3^o = 0$, the payoffs for players 1 and 2 are given by $x_1^o = v_{13} + \alpha\Delta$ and $x_2^o = v_{23} + (1-\alpha)\Delta, 0 \le \alpha \le 1$. Specific values can be obtained from Figure 2.5 by using a ruler.

Hence we have proved the following:

Proposition 2.2 For any 0-normalized non-triangular game $\Gamma = (N, v)$ the following always holds true.

$$x_1^o = v_{13} + \alpha\Delta$$

$$x_2^o = v_{23} + (1-\alpha)\Delta \qquad (2.38)$$

$$x_3^o = 0$$

$$\Delta = \mu(\overline{1b}) - \mu(\overline{1a}), \quad 0 \le \alpha \le 1$$

Remark 2.1 In a triangular cooperative game, the maximum bid a player can expect to get by auctioning his cooperation is clearly the maximum amount he can claim and protect in all possible conditions.

To explain the remark above, suppose coalition {1, 2} is to form and the distribution 50 for player 1 and 30 for player 2 is under consideration by both players. Clearly, they have to agree on it in order to join together. Suppose player 1 wants more than 50 and demands from player 2 an amount $0 < e < 30$ to obtain $50 + e$ while player 2 gets $30 - e$. Player 2 can argue that such demand is not a wise move because the only possible way player 1 could obtain that amount elsewhere is by using his only bargaining alternative, namely coalition {1,3} where he would get $50+e$ provided player 3 gets $20-e$. In such case he (player 2) could protect his 30 by offering player 3 the amount 20 which player 1 cannot match except by giving up his unjustified demand. The same type of argument as the one given above applies to any of the other two distributions of characteristic function value of the coalitions given by the expressions in (2.33).

Remark 2.2 The contingent systems of extended imputations in (2.33) and (2.34) constitute each, a set of maximum sustainable claims and the only conceivable equilibrium in the coalition formation process that may occur previous to the formation of the grand coalition. In fact, the three coalitions may form and the final play of the game represented by the distributions in (2.33) and (2,34) have the same characteristics that von Neumann and Morgenstern identified for the *non-discriminating* *"objective" solution* for the essential zero-sum three-person game, namely:

- The three distributions in (2.33) and (2,34) encompass all strategic possibilities of the game.
- No distribution by itself, but the three of them as an interrelated system, constitutes a solution to the questions: which coalitions may form previous to considering the formation of the grand coalition? And how the proceeds of the coalitions are to be distributed between its members?
- The three distributions are stable in the sense that no equilibrium can be found outside of them and any negotiation between the players can be reasonably expected to conduce to one of these three distributions.
- The stability identified in each case is only a property of the three distributions viewed together. By itself, any one of them can be disrupted by a realizable distribution that could give higher amounts to a sufficient number of players in an alternative coalition.

We may have an equivalent consolidated representation of the contingent systems (2.33).and (2.34). The extended imputation $x^o = (x_1^o, x_2^o, x_k^o)$ may be taken as the composition of three separate occurrences viewed as one. Thus, each system may be equivalently described as the pair $[x^o, \mathbb{C}]$ where x^o is the extended imputation of maximum sustainable claims and $\mathbb{C} = \{C_1, C_2, C_3\}$ is the collection of subsets of N = {1, 2, 3} referred here as a *cover of N* that constitute a support for x^o. That is,

$$[x^o, \mathbb{C}] = [(x_1^o, x_2^o, x_k^o), \{C_1, C_2, C_3\},$$

$$C_1 = \{1,2\}, \ C_2 = \{1,3\}, \ C_3 = \{2,3\} \qquad (2.36)$$

$$\text{and } x^o(C_i) = v(C_i), \ i = 1, 2, 3$$

The pair $[x^o, \mathbb{C}]$ given above and the equivalent contingent system given in (2.33) or (2.34) will be referred to as the *fundamental equilibrium* for the 3-person cooperative game.

Example 2.1 Let us consider the game in Example 1.2. There, $v_{12} = 80, \ v_{13} = 70, v_{23} = 50$.

Replacing these values in (2.11) we obtain $x_1^o = 50, \ x_2^o = 30, \ x_3^o = 20$ Thus, the fundamental equilibrium of this game is given by the contingent system of extended imputations in (2.12) which here becomes

	x_i	x_j	x_k		

Coalition Structure

(50	30	0)	if	{1,2} {3}	forms
(50	0	20)	if	{1,3} {2}	forms
(0	30	20)	if	{2,3} {1}	forms

Or replacing in (2.36) we obtain

$$[x^o, \mathbb{C}] = [(50, 30, 20), \{\{1,2\}, \{1,3\}, \{2,3\}\}]$$

Proposition 2.3 In a triangular game, the maximum bid a player can expect to get is equal to the remainder that the player obtains when bidding for any other player's cooperation.

Proposition 2.4 A game is triangular if an only if the all reminders with respect to the maximum bids are non-negative.

Proposition 2.4 The fundamental equilibrium for a triangular cooperate game is unique

So far, we have established through an auction procedure, the existence and uniqueness of a fundamental equilibrium for triangular games.

It becomes evident, then, that our auction procedure gives us the maximum claims that the players can expect to get and sustain in all possible conditions and that these claims are realizable by any two players that decide to cooperate. The obtaining of this system of claims, supported by the corresponding coalitions, constitutes equilibrium set of possibilities. All three coalitions are on equal footing and any but only one of them may form at a time. Thus, this procedure gives us an answer to the question of which coalitions are likely to form if the grand coalition doesn't form or previous to its formation, it also answers the question of how the proceeds of the coalition that forms is to divide the obtained value.

Remark 2.3 Our auction procedure doesn't provide any clue for determining the likelihood of coalition formation for this would require a way of assigning probabilities to the occurrence of each of the coalitions. However, since cooperation entails joint-moves, joint- strategies and joint-mixed-strategies, consensus probabilities may be determined so as to maximize the expected utility for the players. This way we can give a complete answer to the question of how well the players may do previous to the formation of the grand coalition and we can use this results cope with the insufficiencies of the core as a solution concept. (See chapter on the likelihood of coalition formation)

Remark 2.4 We established by using a geometric interpretation of the characteristic function of a triangular games we can determine graphically the fundamental equilibrium of the game using ruler and compass constructions.

Proposition 2.5 The value level of the fundamental equilibrium extended imputation $x°$ for the cooperative game $\Gamma = (N, v)$ determines the von Neumann number $|\Gamma|$ and conversely

That is $x°$ is a fundamental equilibrium if and only if the von Neumann number of the game $|\Gamma| = \min_{S \subset N} e(x, S)$.

The proof of proposition 2.1 requires LP

- Clearly, independent treatment is given to these subjects whenever we begin with the assumption that a given coalition structure whether the grand coalition or any other one has occurred.

- We consider our approach more in consonance with von Neumann and Morgenstern original work as in the proof on the existence of the core given in the preceding section. This demonstration is based on the concept of extended imputations and a related theorem on the corresponding excess, both introduced in vN-M's book.

- RE-THINKING THE CORE
- THE STRONGER PLAYER PARADOX
- THE CORE OF THE CORE

If the grand coalition doesn't form, then necessarily a coalition structure as in (1.1) for the case of n=3 (a set partition of N different from N) has to occur and for such cases the final payoffs are given as individually rational payoff configurations that consist of a coalition structure with the corresponding distributions of the proceeds of the coalitions in the partition under consideration. However, such theory doesn't address the question of which of the partitions is likely to form but gives stability conditions for payoffs of any coalition structure if it forms..

3. Utility Transfer Analysis and Stability in Bargaining Scenarios

Coalition formation has been and continues to be one of the major research topics in game theory. Its relevance and importance for the economic, politic, social, and behavioral sciences can hardly be overemphasized: business and war strategic alliances, labor unions, cartels, political parties, herd behavior and many other manifestations of group behavior are patterned if not governed by the underlying principles of coalition formation. As most issues in game theory, its study generally has fallen into one of the two main branches in which the theory has been divided after his originally integrated form was introduced by von Neumann and Morgenstern under the name of *"theory of games strategy"*: the cooperative and the non-cooperative. Here, we want to circumvent such dichotomy, first by mentioning how our heuristic construction so far carried out in the preceding section might relate to such approaches and then by presenting a mathematical model of payoffs stability in well-defined bargaining scenarios in which a cooperative game might decompose.

On one hand, we have obtained specific results from our auction approach to coalition formation. Since William Vickrey's pioneer analysis (1961), auction design has received considerable attention in non-cooperative game theory. Japanese auctions are known to be equivalent to *second- price sealed-bid auctions* (Vickrey Auctions). For these auctions it can be shown that true-telling is a dominant strategy. Thus, we may say that the maximum bids we have obtained so far, constitute a Nash equilibrium valuation of the player's cooperation in a stage previous to the formation of the grand coalition. On the other hand, from the point of view of cooperative game theory, the identified maximum sustainable claims which the players can maintain in all conditions and may receive as payment for their cooperation from prospective partners can be shown to constitute as a stable set in a von Neumann and Morgenstern extended sense of a solution for the game.

It becomes evident then, that we have sufficient referents for attempting to connect and make explicit the hidden relations between the two traditionally separated branches of game theory. Notwithstanding and firstly, we want to continue with our constructive approach within a systems perspective and take it to its last consequences. We intend to present an alternative integrated mathematical model free from the conceptual interpretations that originated the two divergent paths in which game theory unfolded after its unified creation. For now, we will concentrate our efforts in characterizing the coalition formation process and the necessary and sufficient conditions for stability of the payoffs around which cooperation may take place from the point of view of utility transfers in well-defined bargaining scenarios. Only after this endeavor is accomplished, we will attempt to undertake the ambitious task of making explicit the relations that that may lay the path and the foundations towards a unified theory of games of strategy.

3.1 Preliminaries

In the preceding section, we identified the maximum sustainable claims which players may get whenever a particular coalition is formed. Also, in the process of obtaining the only conceivable system of stable payoffs that these claims may constitute, we were able to perceive the role that alternative coalitions play in defining the final outcomes. In this section we want to introduce the necessary basic concepts and give a preliminary heuristic description of the interrelated stability conditions that characterize the stable outcomes we encountered in the auction processes. We want to keep an integrated view of conflict and cooperation without falling in the dichotomy of cooperative and non-cooperative interpretations of game theory.

Consider the triangular game of Example 2.1. There we have $v_{12} = 80$, $v_{13} = 70$ and $v_{23} = 50$. Suppose player 3 proposes to player 1 the distribution x = (10, 0, 60) in forming coalition {1, 3}. Further, suppose that player 2 has an ongoing proposal to player 1 that consists on the distribution z = (30, 50, 0) in forming coalition {1, 2}. Clearly, the distribution z poses a real threat to the formation of coalition {1, 3} around distribution x. Player 1 is definitively going to prefer distribution z in {1, 2} to distribution x in {1, 3}. Simply, because 30 >10.   This scenario suggests the necessity of making compensations[22] among players. For player 3 to be able to retain player 1 and save the formation of coalition {1, 3}, a compensation to player 1 in the form of a utility transfer $\varepsilon > 0$ has to be made from his proposed x = (10, 0, 40) to obtain y = (10+ε, 0, 60-ε), where ε must be at least $\varepsilon = 30 - 10 = 20$ to compensate for the gap between z_1 and x_1 and thus to neutralize the threat to coalition {1, 3}. We may also express such utility transfer as an extended imputation. So that, y = x + ξ, or $\xi = y - x = (\varepsilon, 0, -\varepsilon)$. Summing up, in the scenario just described, a utility transfer is found to be necessary to make stable extended imputation x in forming {1, 3} relative to z in in forming {1, 2}. Such relative stability is assured provided $\varepsilon \geq 20$. Only then, $y_1 = z_1$ and when considering y = (30, 0, 20) versus z = (30, 50, 0), player 1 wouldn't have any immediate reason to disrupt coalition {1, 3}.

Going back to our identified fundamental equilibrium we realize that the just mentioned equalizing condition for

---

[22] For an introduction on how final outcomes depend on alternative coalitions and the need for utility transfers as compensations in defining the final apportionment of the value of coalitions, see vN-M (1944) pp.35-36.

the payoffs to player 1 in both coalitions {1, 3} and {1, 2} is fulfilled in all extended imputations that are part of the contingent system given by (2.12) or in its equivalent representation as an extended imputation $x° = (x°_1, x°_2, x°_3)$ with a cover support given by $\mathcal{C} = \{\{1,2\},\{1,3\},\{2,3\}\}$ such that $x(C) = v(C)$ for all C in $\mathcal{C}$. The last conditions $x_1^o + x_2^o = v_{12}$, $x_1^o + x_3^o = v_{13}$ and $x_2^o + x_3^o = v_{23}$ together suggest that our fundamental equilibrium (2.12) can be obtained by solving the simultaneous system of equations given by:

$$1x_1 + 1x_2 + 0x_3 = v_{12}$$
$$1x_1 + 0x_2 + 1x_3 = v_{13} \quad (3.1)$$
$$1x_1 + 1x_2 + 0x_3 = v_{23}$$

Or in equivalent matrix form

$$\begin{pmatrix} 1 & 1 & 0 \\ 1 & 0 & 1 \\ 0 & 1 & 1 \end{pmatrix} \begin{pmatrix} x_1 \\ x_1 \\ x_1 \end{pmatrix} = \begin{pmatrix} v_{12} \\ v_{13} \\ v_{22} \end{pmatrix} \quad (3.2)$$

The corresponding solution is given by

$$\begin{pmatrix} x_1^o \\ x_2^o \\ x_3^o \end{pmatrix} = \begin{pmatrix} 1 & 1 & 0 \\ 1 & 0 & 1 \\ 0 & 1 & 1 \end{pmatrix}^{-1} \begin{pmatrix} v_{12} \\ v_{13} \\ v_{22} \end{pmatrix} = \frac{1}{2} \begin{pmatrix} 1 & 1 & -1 \\ 1 & -1 & 1 \\ -1 & 1 & 1 \end{pmatrix} \begin{pmatrix} v_{12} \\ v_{13} \\ v_{22} \end{pmatrix}$$

Or equivalently

$$\begin{pmatrix} x_1^o \\ x_2^o \\ x_3^o \end{pmatrix} = \frac{1}{2} \begin{pmatrix} v_{12} + v_{13} - v_{23} \\ v_{12} - v_{13} + v_{23} \\ -v_{12} + v_{13} + v_{23} \end{pmatrix} \qquad (3.3)$$

This takes us back to the maximum bids obtained by the players in our auction process and given in (2.11).

Furthermore, we can show that the Japanese auction incremental bidding process of the preceding section can be equivalently described by means of utility transfers and eventually by means of systems of equations in nonnegative variables. To see this, let us retake the threat posed by distribution z = (30, 50, 0) to the formation of coalition {1, 3} around x = (10, 0, 60). Suppose that player 2 has not made this threat explicit jet, as it would be the case when the bidding level is $\beta = 10$, but he is offering to player 1 the same amount that player 3 is offering. That is, w = (10, 70, 0). Then, the utility transfer that takes the bargaining process from y to z is of the form

$$\xi(from\ 2\ to1) = z - w = (\varepsilon, -\varepsilon, 0), \ \varepsilon = 20$$

Such threat to be neutralized requires a transfer

$$\xi(from\ 3\ to1) = y - x = (\varepsilon, 0, -\varepsilon), \ \varepsilon = 20$$

We may describe these two independent moves in a consolidated form. Both the payoffs and the transfers are composed into one payoff and one transfer in an integrated system that allows for viewing as one two separate occurrences[23].

The composed transfer becomes

$$\hat{\xi} = (\varepsilon, -\varepsilon, -\varepsilon), \quad \varepsilon = 20$$

And the composed extended imputations become $\hat{x} = (10, 70, 60)$ and $\hat{y} = (30, 50, 40)$ so that $\hat{\xi} = \hat{y} - \hat{x}$. Both $\hat{x}$ and $\hat{y}$ are solutions to the system of equations

$$1x_1 + 1x_2 + 0x_3 = v_{12} = 80$$

$$1x_1 + 0x_2 + 1x_3 = v_{13} = 70$$

And $\hat{\xi}$ is solution to the system

$$1\varepsilon_1 + 1e_2 + 0\varepsilon_3 = 0$$

$$1\varepsilon_1 + 0\varepsilon_2 + 1\varepsilon_3 = 0$$

It becomes evident that the utility transfer $\hat{\xi}$ with $\varepsilon = 20$ in bargaining from $x_1 = 10$ to $y_1 = x_1 + \varepsilon = 30$ is

---

[23] This omniscient like approach is a characteristic of systems thinking and is the one used in the concept of composition of imputations in von Newman and Morgenstern (1944) p. 360.

equivalent to raising the bidding level β from 10 to 30 in competing for the cooperation of player 1.

Now we will provide a formal mathematical characterization of the above type of bargaining process to obtain necessary sufficient conditions for existence and uniqueness of our fundamental equilibrium. We would like to extend these results to the general n-person cooperative game with $n \geq 3$. For this, as mentioned above, we will approach the bargaining process through a model based on utility transfers which allows us to characterize payoffs stability in well-defined bargaining scenarios. Notwithstanding, most of the examples and heuristics will continue to focus on the 3-person game.

Formally we have:

Definition 3.1 *A cover collection of N, or simply, a **cover of** N is a set* $\mathcal{C} = \{C_1, \ldots, C_k\}$ of subsets of N such that for all j $\in$ N there is $C_l \in \mathbb{C}$ such that j$\in C_l$.

Remark 3.1 The quantity $|\mathcal{C}| = k$ denotes the *cardinality of* the cover $\mathcal{C}$. And clearly, $\bigcup_{i=1}^{k} C_i = N$. The concept of a cover of N generalizes the one of *partition of N*. A partition is a cover with the additional property that for any two sets $C_i, C_j \in \mathcal{C}$, $i \neq j$ $C_i \cap C_j = \emptyset$.

Definition 3.2 A coalition $C \subseteq N$ is said to be *effective* for an extended imputation x if and only if $x(C) = \sum_{j \in C} x_j = v(C)$. A payoff vector x is said to be *C-feasible* if there exists a coalition $C \subseteq N$ which is effective for the extended imputation x.

Definition 3.3 An extended imputation x is said to be *partly-realizable* through coalition C if and only if C is effective for x; or equivalently if and only if x is C-feasible. A cover $\mathcal{C} = \{C_1, \ldots, C_k\}$ of N is said to be an *effective cover* for x, or simply *a cover support* for x, if and only if $x(C_i) = v(C_i)$ for all $C_i \in \mathcal{C}$. That is, if and only if x is C-feasible for all $C \in \mathcal{C}$. Imputations are clearly N-feasible extended imputations. That is, extended imputations realizable through N. The coalition N is both a partition and a cover support for N. If the cover support of an extended imputation x is a partition then we say that the extended imputation x is *fully-realizable.*

Remark 3.2 Extended Imputations which are imputations and in general payoff configurations that have a partition as cover support are clearly fully-realizable and can associated with outcomes that may occur with certainty. However, whenever the cover support of one or more extended imputations is not a partition then the extended imputations supported by the cover are not fully-realizable but partly-realizable for all the coalitions in the cover. This partial realizability allows us to describe multiple contingent outcomes viewed as one omniscient reality which is a basic characteristic of our fundamental equilibrium. That is, an equilibrium not of actualities but one of possibilities that may describe occurrences with certainty as a special case.

Definition 3.4 An extended imputation x is said to be *attainable* if and only if there exist a cover C of N that is effective for x.

In general, attainable extended imputations are realizable in a conditional sense. That is, subject to the formation of a coalition in the cover. We note that except in the case of partitions or sub partitions, the occurrence of one coalition excludes the occurrence of other coalitions in the cover. Then, we say that attainable extended imputations are *contingent–realizable.*

It becomes clear then that extended imputations with cover support may describe contingent- realizable payoffs whenever the cover is not a partition and describes fully-realizable payoffs when the cover support happens to be a partition.

Example 3.1 The following table gives us a sample of some collections of subsets of N and the corresponding Characteristic matrices:

Collection of subsets S	$\|S\|$	Characteristic matrix W

$S_1 = \{\{1, 2\}, \{2, 3\}\}$ 2

$W_{(2x3)} = \begin{pmatrix} 1 & 1 & 0 \\ 0 & 1 & 1 \end{pmatrix}$

$S_2 = \{\{1, 2, 3\}\}$ 1

$W_{(2x3)} = (1 \quad 1 \quad 1) = J^t$

$S_3 = \{\{1, 2\}, \{1, 3\}, \{2, 3\}\}$ 3

$W_{(3x3)} = \begin{pmatrix} 1 & 1 & 0 \\ 1 & 0 & 1 \\ 0 & 1 & 1 \end{pmatrix}$

$S_4 = \{\{1, 2\}, \{3\}\}$ 2

$W_{(2x3)} = \begin{pmatrix} 1 & 1 & 0 \\ 0 & 0 & 1 \end{pmatrix}$

$S_5 = \{\{1, 3\}\}$ 1

$W_{(1x3)} = (1 \quad 0 \quad 1)$

Of the collections of coalitions listed in the table above, only the first four are covers of N. Note that for the game of Example 2.1, the collection S_1 is a cover support for the extended imputation x = (30, 50, 40) but S_2, S_3 and S_4 are not. Also, let us observe that x is realizable through coalition {1, 2} in S_4 but it is not attainable since it is not realizable for coalition {1}. Similarly, x is realizable for coalition {1, 3} in S_5 but is not attainable because the collection S_5 is not a cover.

Definition 3.5 An extended imputation z **dominates** another extended imputation y if and only if there exist a coalition $S \subseteq N$ and an S-feasible extended imputation x such that $x(S) \geq z(S)$ and $z_j > y_j$ for all j in S. Whenever z dominates y through coalition S, we write $z \succ_S y$.

Note that S-feasible extended imputations cannot be dominated via coalition S and that detached extended imputations (se Definition 2.x) are dominated by no extended imputation. In what follows, we will be pay special attention to the set of detached extended imputations that have cover support. That is the set of attainable detached extended imputations that will define the bargaining scenarios we want to focus on to determine conditions that characterize the systemic stability we identify through our constructive procedure for modeling the process of coalition formation as an auction.

Remark 3.3 We note that a collection $C = \{C_1, \ldots, C_k\}$ is a cover support for an extended imputation x if and only if C is a cover of N and x is C-feasible for all C in C Attainable extended imputations are not necessarily realizable for all its components at once but are C-realizable for all coalitions in the cover. If we let the vector v in R^k denote *the restriction of the characteristic function values* to the coalitions in the cover C then, we may say that the cover C of N is a cover support for the extended imputation x if and only if

$$Wx = v, \qquad (3.4)$$

Here W is the characteristic matrix of the cover C and v is the vector of corresponding coalition's values

Definition 3.6 A *utility transfer* is a vector $\xi \in R^n$, $\xi = y - x$ where x and y are extended imputations.

Utility transfer vectors summarize the transfers of utility that may take place whenever players are bargaining in relation to a given payoff vector x, and as a result, the payoff vector $y = x + \xi$ is obtained.

Definition 3.7 if x and y are two different extended imputations with a common cover support $\mathcal{C}$, then the vector $\xi = y - x$ is said to be a non-null *admissible utility transfer* by the cover $\mathcal{C}$.

Definition 3.8 Relative to a cover $\mathcal{C}$ of N, a non-null admissible utility transfer is said to be *Pareto-efficient* if and only if $J^t \xi = 0$.

Note that Pareto-efficient transfers are the type of transfers that take place in Pareto-efficient distributions. That is, distributions where any player's gain in a transfer must come from one or more than one player's losses. Thus clearly, the concept of set of extended imputations with cover support generalizes the one of imputations. In this context, imputations can be defined as extended imputations supported by the collection $\mathcal{C} = \{N\}$.

Remark 3.3 Clearly, a non-null utility transfer ($\xi \neq 0$) is admissible by a cover $\mathcal{C}$ of N if and only if $W\xi = 0$. This becomes evident since $Wx = Wy = v$ and $W\xi = W(y - x) = 0$. Note that the existence of an admissible non-null

Pareto–efficient utility transfer by a cover $\mathcal{C}$ of N, indicates that the augmented system $W\xi = 0$, $J^t\xi = 0$ has a non-trivial solution. This happens whenever the rank of W is less than n and the characteristic vector of coalition N is linearly dependent[24] of the characteristic vectors of the coalitions in the cover $\mathcal{C}$.

Definition 3.9 A non-null utility transfer ($\xi \neq 0$) for a cover $\mathcal{C}$ of N is said to be *zero-sum* if and only if the following two conditions are simultaneously satisfied:

$$W\xi = 0, \text{ (admissibility)} \quad (3.5)$$

$$J^t\xi = 0 \text{ (Pareto- efficiency)} \quad (3.6)$$

Clearly, if $\xi \neq 0$ is admissible by a cover $\mathcal{C}$ of N, so is $-\xi$ since $W\xi = 0 = W(-\xi)$.

In the following example we will observe how the definitions and basic concepts introduced above provide us with the necessary concepts to model the interpersonal bargaining among the players that may take place in the process of coalition formation around possible binding agreements. We will see how utility transfers may be

---

[24] See Appendix on Mathematical Basis: Linear Algebra

demanded and justified using alternative coalitions that can provide better outcomes to a sufficient number of players. Of the utility transfers that may take place only one of two possible directions appears to be enforceable. From the point of view of inter-player bargaining dynamics we expect the utility transfers, within the cover under consideration, to be directed from players with sufficient bargaining alternatives towards players without them. That is the direction of the admissible transfers in a given scenario must reflect the incremental cost trend emerging from the competition between two players for the cooperation of the third one.

Example 3.2 (a) (Buyer's / Seller's Market Archetype) Consider the 3-person game, $v(\{1,2\}) = a$ $v(\{1,3\}) = b$, $a \geq b > 0$, and $v(S) = 0$ otherwise. This game is clearly a non-triangular game. Note that though the coalition of all players can form and obtain the value a, here, we want to continue our inquiry by focusing in a stage previous to considering the formation of the grand coalition. The graphic solution and corresponding claims obtained by appropriate replacing in (2.23) is given in Figure 3.1 bellow:

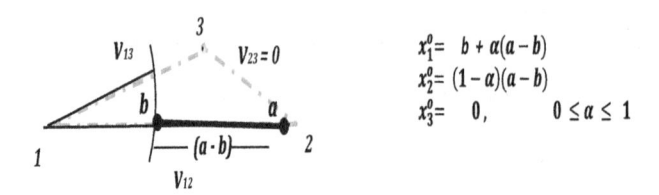

Figure 3.1 Graphic for maximum claims in seller's market

In this game, two buyers, players 2 and 3 are interested in buying a commodity from a seller, player 1. The quantity a represents the value created if player 2 buys from player 1. That is, the amount a, is the difference between the maximum value that the buyer, player 2, is willing to pay for the commodity in consideration and the minimum value to be accepted by the seller, player 1. Similarly, the quantity b is the difference between the worth of the commodity to player 3 and the minimum acceptable selling value to the seller. Thus, the bargaining between players in both cases is for the value created by each possible 2-person coalition.

Now, we want to model the bargaining process captured in the three independent auctions in chapter 2 but as an integrated system by using our utility transfer analysis.

At an initial stage, except for the zero-disagreement individual value supported by each one-person coalition, the only coalitions that may be effective for bargaining purposes are the ones in the cover $C = \{\{1, 2\}, \{1,3\}\}$ with characteristic matrix $W = \begin{pmatrix} 1 & 1 & 0 \\ 1 & 0 & 1 \end{pmatrix}$ and admissible utility transfer $\xi = (\varepsilon, -\varepsilon, -\varepsilon)^t, \varepsilon > 0$. Note de direction of the transfer is towards increasing the value for player 1 and is characterized by the fact that $J^t\xi = -\varepsilon < 0$. Also, we have $W\xi = 0$ indicating that the transfer is admissible. It will be seen below (proposition 3.2), this structure by itself to be clearly unstable given that it reflects the advantage of the seller in seller's market.

In considering an extended imputation $x = (x_1, x_2, x_3)^t$ supported by cover $C = \{\{1, 2\}, \{1, 3\}\}$ of N, with $0 < x_1 < a$, player 1 may ask player 2 for a utility transfer, using as leverage his alternative to go with player 3. Similarly, he may demand additional utility from player 3, by arguing his alternative to join with player 2. On the other hand, it is clear that the utility transfer $-\xi =$

$(-\varepsilon,\ \varepsilon,\ \varepsilon)^t, \varepsilon > 0$ is not enforceable; players 1 and 3 have no bargaining alternatives and hence cannot enforce the utility transfer $-\xi$ in an interpersonal bargaining context. To enforce some kind of demand if any, would require for players 2 and 3 to coordinate a counter strategy by acting as one i.e., to form a union or syndicate, transforming the original game in a different one of pure bargaining between two players. Namely, player 1and the formed syndicate as the other player. The identified possibility of such type of coalition formation has to be considered to exhaust all possibilities. Such behavior is the equivalent to having a bidder's collusion in the auction process and will be address later as an important behavioral pattern in defining the possible final outcomes of a negotiated rather that arbitrated cooperative game.

For a = 70 and b = 50 a graphic model of the bargaining scenario for the 3-person market archetype consisting of either, two buyers and one seller or of two sellers and one buyer, is given in Figure 3.2 below. Note that in this game, there are only two possible scenarios where bargaining interaction among players may take place. One supported by the cover $\mathcal{C} = \{\{1, 2\}, \{1,3\}\}$ that consist of all extended imputations between the points A and B. And the other scenario is the one supported by the cover $\hat{C}$ = {{1,2},{3}} that consist on all the extended imputations between the points B and C (a mathematical well-defined concept of bargaining scenario will be given subsequently).

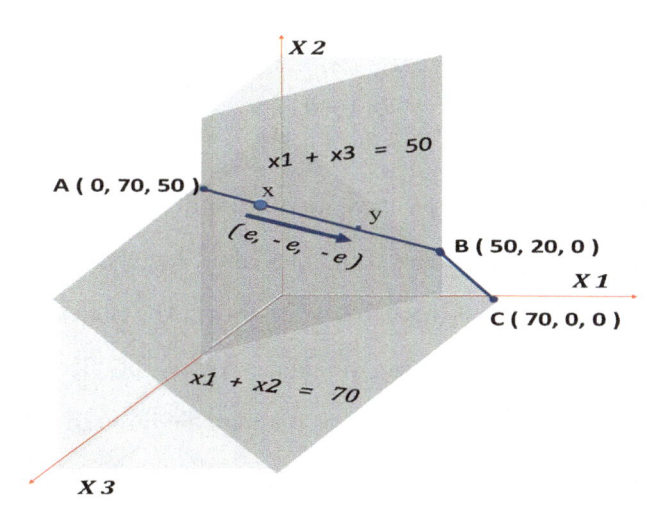

Figure 3.2 Enforceable utility transfers in the 3-person simple market archetype

Note also that the vector difference B - A = (50, -50, -50) reflects the general structure of the admissible utility transfer $\xi = y - x = (\varepsilon, -\varepsilon, -\varepsilon)^t$. It indicates for $\varepsilon > 0$, the only enforceable direction in the given scenario reflecting the fact that players 2 and 3 have to bid competitively for the cooperation of player 1. Clearly, both x and y are extended imputations supported by the cover $\mathcal{C} = \{\{1, 2\}, \{1,3\}\}$ and ξ is a utility transfer vector admissible by the cover $\mathcal{C}$.

For the specific values a=70 and b=50 the claims for the players given in Figure 3.1 become:

$$x_1^o = 50 + 20\alpha$$
$$x_2^o = 20(1 - \alpha)$$
$$x_3^o = 0, \ 0 \leq \alpha \leq 1$$

Remark 3.4 By now we have the picture that bargaining scenarios emerge in the form of claims together with alternative coalitions that can support them. The evidently emerging bargaining scenario here, consist of all the extended imputations from A to B attainable through the cover $\mathcal{C}$. Clearly, the final possible outcomes where the bargaining process in this scenario converges is to the extended imputation B = (50, 20, 0) which summarizes the conditional system of contingent payoffs: (50, 20, 0) if coalition {1,2} forms and (50, 0, 0) if coalition {1,3} forms. An initial bargaining departure point for player 1 and 2 could be any point between A and B, say x = (10, 60, 40)^t.

Another intermediate bargaining point would be y = (30, 40, 20)^t. The bargaining pairs [x, $\mathcal{C}$] and [y, $\mathcal{C}$] represent the contingent systems:

(i) $[x, \mathcal{C}] \sim \begin{bmatrix} (10, 60, 0) \\ (10, 0, 40) \end{bmatrix}$ and (ii) $[y, \mathcal{C}] \sim \begin{bmatrix} (30, 40, 0) \\ (30, 0, 20) \end{bmatrix}$

Clearly the extended imputation (10, 60, 0)^t is dominated by (30, 0, 20)^t via coalition {1,3} and the extended imputation (10, 0, 40)^t is dominated by (30, 40, 0)^t via coalition {1,2}. In each case player 2 and player 3 have to make subsequent compensations in the form of utility transfers to neutralize the threat of an alternative for player 1 of a more attractive offer that have the potential of dissolving the initial proposition. That is, for ε = 20 (30, 40. 0) = (10 +ε, 60-ε, 0) and (30, 0, 20) = (10+ε, 0, 40-ε). Clearly there, y = x+ξ and the only strategically viable direction for ξ is the one where J^t ξ < 0, that is, for ξ =(ε, -ε, -ε), ε > 0.

(b) Taking a step toward generality, for n > 3, consider the cover $\mathcal{C}$ ={{1, 2, 3, 4}, {2, 3, 4, 5}, {1, 4, 5}, {1, 3, 5}} in a 5-person cooperative game the corresponding characteristic matrix is given by:

$$W = \begin{pmatrix} 1 & 1 & 1 & 1 & 0 \\ 0 & 1 & 1 & 1 & 1 \\ 1 & 0 & 0 & 1 & 1 \\ 1 & 0 & 1 & 0 & 1 \end{pmatrix}.$$

The cover under consideration admits utility transfers of the form $\xi = (-\varepsilon, -3\varepsilon, 2\varepsilon, 2\varepsilon, -\varepsilon)$, $\varepsilon > 0$. We note that, $-\xi$ is also admissible. However, $-\xi$ is not enforceable for it would imply player 2 demanding from player 3. Clearly, the former has no bargaining alternative against but strictly depends on the later. In both cases above, the common characteristic of the transfers with enforceable direction is its non-Pareto-efficient characteristic where the utility transfer satisfies the condition $J^t \xi < 0$.

Definition 3.10 A non-null utility transfer admissible by a cover $\mathcal{C}$ of N is said to be *enforceable* (from an interpersonal bargaining point of view) if and only if only if $J^t \xi < 0$.

3.2 Utility transfers in bargaining scenarios

The need for utility transfers as compensations to a certain player when facing the dissolution threat posed by a dominating bargaining alternative which offered the pivotal player a higher utility retribution for his cooperation was demonstrated in remark 2.1 of section 2, for the 3-person market game. Here, will give along this line of thinking the general idea behind the concept of bargaining scenario, defined as a relational system of binding agreements where bargaining alternatives are used to protect player's claims so that these claims have to be equalized in any realizable bargaining alternative within a given scenario.

We may observe that two different coalitions S and T in N bare a direct strategic relation in the game $\Gamma = (N,v)$ if and only if their intersection is not empty. If such is the case then we say that the players in such intersection constitute an *implicit coalition*. Based on this identified new type of "virtual coalition" we will extend the concept of domination among extended imputations to groups of players with parallel interest not explicitly expressed as in the case of coalitions with explicit characteristic function value.

Definition 3.11 Given two extended imputation x and y and two coalitions S and T in N, such that $S \cap T \neq \emptyset$. Suppose x is S-feasible and y is T-feasible, then we say that y dominates x via implicit coalition $D = S \cap T$ if and only if $y_j > x_j$ for all $j \in D$.

Clearly, if y dominates x implicitly, then y is a *direct threat* to the formation of S around distribution x and we are faced with a situation similar to the one described in remark 2.1. In fact, the extended imputation y may be considered a threat to the stability of x only if y(D) > x(D). The amounts x(D) and y(D) are the value levels of the extended imputations x and y relative to the coalitions S and T respectively. In the case the 3-person cooperative game the implicit coalitions contain only one player. As with the definitions of domination between imputations and between extended imputation, vN-M (1944) p.264,p.272-273 and p.367, the conditions below (3.7) will be referred as the preliminary conditions, and (3.8) as the main condition.

(i)   S ∩ T ≠ ∅ and ii) y(D) > x(D)   (3.7)

(iii) $y_j > x_j$ for all j ∈ D   (3.8)

Example 3.3   For the e-essential 7-person 0-normalized game with S = {3, 4, 5, 6, 7} and T = {1, 2, 3, 4, 5} and characteristic function values given by v(S) = 100, v(T) =150, and v(C) = 0 for all other C in N. Here the implicit coalition is S ∩ T = D = {3, 4, 5}. For y = (20, 45 , 35 , 20 , 30

10, 25)^t and x =(16, 14, 10, 15, 25, 20, 30) , clearly x is S-feasible since x(S) = 100 = v(S) and y is T-feasible since y(T) =150 . Clearly (3.7) and (3.8) holds for x and y so that y dominates x via implicit coalition D

Now, we will proceed to give a formal definition of the bargaining scenario concept. Then, we will look for claims in the form of extended imputations that can be maintained in all possible conditions within the given scenario and may be considered stable from the point of view of utility transfers. By doing so, we focus on internally stable systems. Clearly, if every player may obtain the same amount in all conditions under consideration then there is no room for internal domination neither for requesting utility transfers[25]. Heuristically, potential members of a coalition S may consider to join together by dividing the proceeds v(S), say implementing a *joint strategy* x^S in proportion to each player's marginal contribution. That is, player j in S is contemplating to receive $x_j = m_j v(S)/m(S)$. In fact, any initial distribution would serve as departing point as suggested

---

[25] Utility transfers in convergent schemes have been object of studies in the cooperative game theory literature were introduced by Stearns (1968) have been used in Kalai, Maschler, and Owen (1973), in Billera (1976) and in Bennett, Maschler and Zame(1994).

in section 3.1 and observed in example 3.2. Once an initial payoff distribution is proposed, players begin to look for bargaining alternatives. Each player will try to justify his actual claim, will demand a higher amount or in the absence of bargaining support will give up his current standing for a lesser payoff if he is to be admitted in the coalition under consideration. The process of using bargaining alternatives to justify higher claims, invariably leads players to consider alternative coalitions and demanding utility transfers, until one among several equilibrium structures is reached. We will see how, among the equilibrium structures that may emerge, some that will be termed as "complete", consist of linearly independent sets of maximal dimension that admit no utility transfers.

However, instead of entering the above interactive complexity, by explaining the decisions and steps that may lead the players to get involved, in a particular bargaining scenario, we will focus on the bargaining scenario itself. The bargaining scenario will serve then, as the systemic unit of analysis, that allows us to describe what players may claim, though not necessarily maintain, in all possible conditions in the cover under consideration.

Let $X(\mathcal{C}, v)$ be the set of all extended imputations supported by a given cover $C = \{C_1, ..., C_k\}$ of N. That is, $X(\mathcal{C}, v) = \{x \in R_+^n \mid W x = v/C \}$. Here we will simply write $W x = v$, assuming the vector v always denotes the corresponding restriction of the characteristic function to the coalitions in the cover $\mathcal{C}$. Any solution to the system $W x = v$, $x \geq 0$ is an extended imputation n-vector of claims realizable only through the cover $\mathcal{C}$.

The rank of the characteristic matrix W of a cover $\mathcal{C}$ of N depends on the number of linearly independent rows of W. If the characteristic vectors of a collection of subsets of N are linearly independent, we say that the corresponding coalitions are *linearly independent subsets of N*. A cover collection of linearly independent subsets of N is referred to as a *linearly independent cover of N*. Here, without loss of generality we will restrict our attention to linearly independent covers of N so that $C = \{C_1, ..., C_k\}$ with cardinality $|C| = k \leq n$ and the corresponding characteristic matrix $W_{k \times n}$ has rank(W) = k. The linearly independence assumption is made to eliminate both, redundancy and inconsistency from our analysis.

Definition 3.12 A pair [X($\mathcal{C}$, v), $\mathcal{C}$] is a *bargaining scenario* for the game Γ = (N, v) if and only if the cover $\mathcal{C}$ is a linearly independent collection of subsets of N and X($\mathcal{C}$, v) is the set of all extended-imputations attainable through the cover $\mathcal{C}$.

Here, the set X($\mathcal{C}$, v) = {x | W x = v, x ≥ 0} will be referred to as the *attainable set* of the corresponding bargaining scenario and the collection $\mathcal{C}$ of subsets of N as the corresponding linearly independent *cover support*. The constitutive elements of the bargaining scenario above is a pair of the form [x, $\mathcal{C}$] here referred to as a *bargaining pair*, where x is an extended imputation realizable through $\mathcal{C}$. Any bargaining pair [x, $\mathcal{C}$] represents a conditional system of contingent outcomes where, for all coalitions in $\mathcal{C}$, players in a coalition C ∋ $\mathcal{C}$ receive the corresponding payoffs in x^C provided the coalition C forms.

Realizable sets are *polyhedral sets*, since clearly they result from the intersection of a finite number of half spaces. A *polytope* is a polyhedral set that is closed and bounded. Polytopes may also be defined as the convex hull of a finite number of points. The two descriptions can be shown to be equivalent (See appendix on convex sets)

Proposition 3.1 The realizable set $X(\mathcal{C}, v)$ of a bargaining scenario $[X(\mathcal{C}, v), \mathcal{C}]$ is a convex polytope.

Proof: The set-in consideration is clearly a convex polyhedral set since it is the intersection of a finite number of half spaces. Thus, we only need to establish that it is bounded. Let $a_j = \max\limits_{i|j \in C_i} v(C_i)$, j=1,...,n , $C_i \in C$. Clearly, then, for all x in X(C , v) we have $0 \leq x_j \leq a_j$, j = 1,....,n. Hence $\|x\| = \sqrt{\sum_{j=1}^{n} x_j^2} < A = 1 + \sqrt{\sum_{j=1}^{n} a_j^2}$, for all x in $X(\mathcal{C}, v)$ .//

Theorem 3.2 (Structural instability) Let $\mathcal{C}$ be a linearly independent cover of N, with $0 < |\mathcal{C}| = k < n$. Then, there exist always a utility transfer $\xi \neq 0$ admissible by $\mathcal{C}$.

Proof: We may consider this theorem as a particular case of the general one on solutions of linear systems of homogeneous equations. (Note that the homogenous system $W\xi = 0$ has n unknowns and k < n equations and hence must have a non- null solution).

Theorem 3.3 (Complete structural stability) Let C = {C₁,...,Cₙ } be a linearly independent cover collection of subsets of N, with $|C| = n$. Then, there exist no utility transfer $\xi \neq 0$ admissible by the cover C.

Proof: As in Theorem 3.2, we may readily verify this to be a particular case of the theorem on solutions of homogeneous systems of equations.

Corollary 3.1 If a cover $\mathcal{C}$ is linearly independent of maximal dimension, then the realizable set $X(\mathcal{C}, v) = \{x°\}$ is a singleton provided $x° \geq 0$. The dimension of the corresponding polytope is zero.

Proof: Clearly, the rank of W is n. Then, by theorem 3, the homogeneous system has no solution, and hence, the system $W x = v$ has a unique solution given by $x° = W^{-1}v$. If $x° \geq 0$, then $x°$ is an extended imputation, and $X(C, v) = \{x°\}$ is a singleton and hence a polytope of dimension 0.

The type of instability inherent to bargaining scenarios and observed whenever admissible utility transfers are allowed by the structure of the coalitions in a given cover, is termed here as *structural instability*. The specific structural instability that occurs when the admissible utility transfers are Pareto-optimal (definition 2.5), characterizes the type of transfers that may occur, generally, in pure bargaining scenarios and will be referred to, here, as *instability of the Pareto-optimal* type. Such instability may be considered as neutral or symmetric instability. The *Pareto-suboptimal structural instability* is the type of instability that we want to characterize in order to determine the conditions under which such instability cannot occur.

Now, we will proceed to give necessary and sufficient conditions for the structural admissibility of transfers, as an application of the fundamental theorem of linear algebra in the form of *Fredholm [26] alternative for matrices* form, Strang (2009). In our case, it gives us the following theorem.

Theorem 3.4 *(Utility transfer admissibility theorem)* Given a cover $\mathcal{C}$ of N with characteristic matrix W, then either the system (1) $W^t \gamma = J$ has a solution, or the system (2) $W\xi = 0$, $J^t \xi \neq 0$ has a solution. But not both (1) and (2) can have a solution.

---

[26] See Strang (2009)

Corollary 3.2 If the system $W^t \gamma = J$ has a solution and there exist $\xi \neq 0$ such that $W\xi = 0$, then it must be that $J^t \xi = 0$.

Proof: $\xi^t W^t = 0$ and $W^t \gamma = J$ with $\xi \neq 0$ implies $(\xi^t W^t) \gamma = \xi^t (W^t \gamma) = 0$, hence $J^t \xi = 0$.

Definition 3.2 Whenever the system $W^t \gamma = J$ has a solution, the corresponding cover C of N is said to be a *linearly balanced collection of subsets of N,* or simply, *an l–balanced cover of N*.

Thus, Corollary 3.2 states that whenever an *l*-balanced collection admits a utility transfer, it has to be zero-sum. In other words, *l*-balanced collections admit, if any, only Pareto-optimal transfers.

The following definitions of balanced collections discriminate more deeply than those given in Bondareva (1963) and Shapley (1967).

Definition 3.3 A collection $C = \{C_1, \ldots, C_k\}$ of subsets on N with characteristic matrix W, is said to be:

> *l-balanced* if $\exists\, \gamma \in R^k \mid W^t \gamma = J$, γ unrestricted
> (linearly-balanced)

- *nn-balanced* if $\exists\ \gamma \geq 0\ |\ W^t \gamma = J$, (non-negatively-balanced)
  - *p-balanced if* $\gamma > 0$, (positively-balanced)
  - *w-balanced if* $\gamma \geq 0$ *with* $\gamma_i = 0$, for some $i = 1, \ldots, k$ (weakly balanced)
- *n-balanced* if $\exists\ \gamma \in R^k\ |\ W^t \gamma = J$, $\gamma_i < 0$, for some $i=1,\ldots,k$ (negatively balanced)
  - *pn-balanced if* $\nexists\ i = 1, \ldots, k\ |\ \gamma_i = 0$, (only positive and negative coefficients).
  - *wn-balanced if* $\exists\ i = 1, \ldots, k\ |\ \gamma_i = 0$, (positive, negative and zero coefficients).

An *l*-balanced collection C is said to be *minimal* if it has no *l*-balanced proper sub-collection. Clearly, weakly balanced covers, w-balanced or wn-balanced, can't be minimal. The vector $\gamma \in R^k$ here, is referred to as the *balancing vector* of the collection C. The elements γ_i, $i = 1, \ldots, k$ are the *balancing coefficients* of the corresponding sets in $\mathcal{C}$. Note that both types of covers: non-negatively balanced and negatively balanced, are partitioned into subclasses, according to weather the cover has zero balancing coefficients.

Theorem 3.5 A non- weakly l–balanced cover $\mathcal{C} = \{C_1, \ldots, C_k\}$ of N is a linearly independent collection of subsets of N if an only if C is minimally-balanced.

Proof: If C is a linearly dependent collection and minimal l-balanced. Then the characteristic row vectors of the coalitions in $\mathcal{C}$ satisfy both equations (1) $\alpha_1 W_1 + \ldots + \alpha_{k-1} W_{k-1} = W_k$ and (2) $\gamma_1 W_1 + \ldots + \gamma_{k-1} W_{k-1} + \gamma_k W_k = J$. Replacing W_k in the balanced equation with the corresponding linear combination we obtain a balanced sub-collection consisting of k-1 coalitions. This contradicts our assumption that C is minimally l-balanced. Hence, C not linearly independent implies C not minimally balanced, . Conversely, if C is a linearly independent collection and not minimal l-balanced then both (1) $\gamma_1 W_1 + \ldots + \gamma_s W_s + \ldots + \gamma_k W_k = J$ and (2) $\delta_1 W_1 + \ldots + \delta_s W_s = J$ must hold. Subtracting (2) from (1), we conclude that the vectors $W_1 \ldots W_k$ are linearly dependent. This contradicts our assumption. That is, C not minimally balanced implies C not linearly independent.

Here we derive as corollaries the following statements originally proved by Shapley (1967).

Corollary 3.3 The balancing coefficients in a minimal l-balanced collection C are unique and the maximum number of coalitions is equal to n.

Proof: Both assertions follow directly, from the linear independence of the rows of the characteristic matrix of C.

Corollary 3.4 A p-balanced collection of linearly independent sets is minimally-balanced.

Proof: Clearly, a p-balanced cover is also l-balanced. Hence, by theorem 3.5, it must be minimally balanced.

Theorem 3.6 (Pareto-optimal instability of partitions) A non-null utility transfer admissible by a partition is zero-sum.

Proof: If the cardinality of the partition C is n, then the corresponding characteristic matrix is the identity matrix. And clearly the cover C doesn't admit any non-null utility transfers. Now, any partition with $|C| < n$, by theorem 3.2, admits always non-null utility transfers. Since every partition is minimally p-balanced by corollary 3.2 the admissible transfers must be Pareto-optimal and hence, these fulfill the conditions of definition 2.6 for zero sum transfers.

Example 3.4 Consider the 3-person zero-sum game with characteristic function given by $v(S) = 1$ if $|S| = 2$ or 3 and $v(S) = 0$ otherwise.

Here, the bargaining scenarios are basically composed of 3-person market archetypes. In each scenario, enforceable utility transfers may take place until the players that are affected adversely, make use of a bargaining alternative to protect their claims, against further demands from the player in the dominant position (the one with bargaining alternatives).

We may observe, in figure 3.1, the utility transfers in all possible bargaining scenarios with coalitionally rational extended imputations, converge to the equilibrium described by von Neumann and Morgenstern in the non-discriminatory solution obtained for this game.

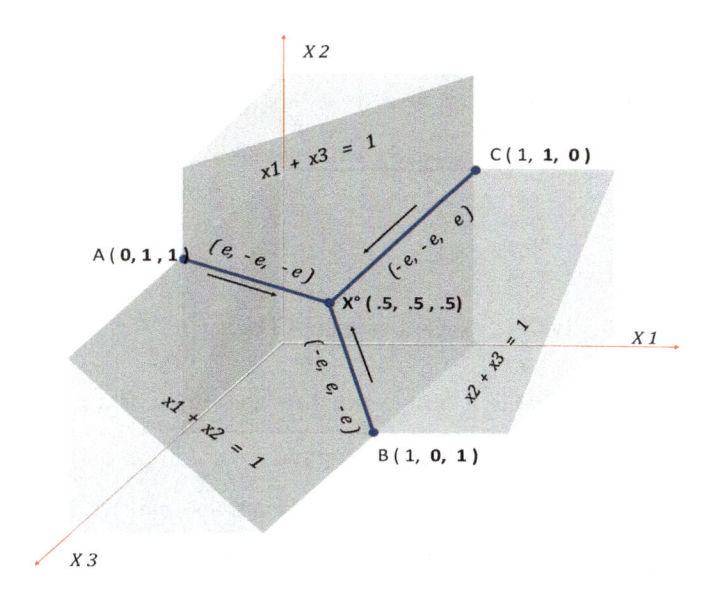

Figure 3.3 Enforceable utility transfer structure for the 3-person zero-sum game.

The extended imputation $x° = (0.5, 0.5, 0.5)^t$ supported by the cover collection $C' = \{\{1, 2\}, \{1, 3\}, \{2, 3\}\}$, when understood as a system of possibilities, appears as a bargaining attractor where all negotiations in the corresponding bargaining scenarios will end up taking place. In this particular case, the emergent equilibrium coincides with von Neumann and Morgenstern "objective" solution as described by system (1) in the introduction of this work.

In figure 3.3 above, we may observe also, that the utility transfer structure (e, -e, -e), is the one associated to the bargaining scenario with realizable set

$X(C, v) = \{x \mid Wx = v, x \geq 0\}$ where $x = [x_1\ x_2\ x_3]^t \geq 0$, $C = \{\{1,2\},\{1,3\}\}$,

$W = \begin{bmatrix} 1 & 1 & 0 \\ 1 & 0 & 1 \end{bmatrix}$ and $v = \begin{bmatrix} 1 \\ 1 \end{bmatrix}$.

Explicitly, $X(C, v) = \{x \mid x = \alpha \begin{bmatrix} 0 \\ 1 \\ 1 \end{bmatrix} + (1 - \alpha) \begin{bmatrix} 1 \\ 0 \\ 0 \end{bmatrix},\ 0 \leq \alpha \leq 1\}$.

That is, a convex polytope described as the convex combination of its extreme points.

Without bargaining alternatives, players 2 and 3 are subject to be driven from A (0,1,1) towards Q(1,0,0). The segment AX° is a subset of the segment AQ (the point Q is below the plane defined by $x_2 + x_3 = 1$). In figure 3. when approaching x° from A with utility transfers of the form (e,-e,-e) Players 2 and 3 may put an end to these utility transfers precisely at x° (0.5, 0.5, 0.5) by using as bargaining alternative coalition {2, 3} with v({2, 3}= 1.